PRENTICE HALL

# SCIENCE EXPLORER

# *Astronomy*

## Teaching Resources with Color Transparencies

PRENTICE HALL
Needham, Massachusetts
Upper Saddle River, New Jersey

## Teaching Support

| | |
|---|---|
| **Lesson Plan** (1 per section) | Streamline your path through every section with a core set of activities and teaching suggestions |
| **Color Transparencies** (avg. 4 per chapter) | Visualize key concepts dynamically with engaging art from every chapter |
| **Complete Answer Keys** | Find all the answers in one section at the back of the book |

## Reading & Review

| | |
|---|---|
| **Section Summaries** | Encourage all readers and limited English proficiency (LEP) students with readable summaries (1 per section)that match the *Guided Reading Audiotapes (English or Spanish)* |
| **Review & Reinforce Worksheets** (1 per section) | Motivate students to build vocabulary, review main ideas, and interpret diagrams, charts, and graphs |

## Enrichment

| | |
|---|---|
| **Enrich Worksheets** (1 per section) | Encourage all students to read, write, and visualize as they apply core concepts in a new context |

## Labs & Projects

| | |
|---|---|
| **Skills Labs and Real-World Labs** (2 per chapter) | Help students get organized with blackline masters of Student Edition Labs—each is an easy-to-grade format |
| **Chapter Projects** (1 per chapter) | Guide students through the Chapter Projects in the student text with extensive teacher notes, scoring rubrics, and student worksheets with project rules and checklists |

## Assessment

| | |
|---|---|
| **Performance Assessments** (1 per chapter) | Assess student problem-solving and process skills using scoring rubrics and suggested outcomes |
| **Chapter and Book Tests** (1 per chapter and 1 per book) | Prepare students for local and state exams with a variety of questions—all different from the Computer Test Bank |

## Interdisciplinary Exploration

| | |
|---|---|
| **Science, Math, Social Studies, and Language Arts** (1 per book) | Support the Interdisciplinary Exploration in the student text with 5 activities you can teach on your own or use for team teaching |

RESOURCE PRO®

**The Resource Pro® CD-ROM with Planning Express®** is the ultimate scheduling and lesson planning tool. Resource Pro® allows you to preview all the resources in the *Science Explorer* program, organize your chosen materials, and print out any teaching resource. With Planning Express® you can follow the suggested lessons or create your own, using resources from anywhere in the program.

PRENTICE HALL
Needham, Massachusetts
Upper Saddle River, New Jersey



ISBN 0-13-436629-8
1 2 3 4 5 6 7 8 9 10 05 04 03 02 01 00 99 98

## Chapter 1
## Earth, Moon, and Sun



## Chapter 2
## The Solar System

## TABLE OF CONTENTS *(continued)*

CHAPTER 1

# Where's the Moon?

1

The following steps will walk you through the Chapter 1 Project. Use the hints as you guide your students through planning, recording observations of the moon, and class presentations.

- This project can be started during any phase of the moon. However, students may have more success with their first observations if you start between the new moon and the waning full moon. At this stage, students can easily see the moon during early evening hours. Students are more likely to have success and get comfortable with recording their observations. If the study starts during stages of the moon's cycle when the moon is not above the horizon during the early evening hours, remind students that seeing that the moon is not visible on a clear night is an important observation and should be recorded.
- Because students commonly associate the moon with night, they may notice the moon more easily if this project is done during the winter months, when night lasts longer.
- Distribute copies of Chapter 1 Project Worksheets, page 10. Worksheet 1 will help students become familiar with the compass directions at their observation site. All students should record observations on an observation sheet similar to that in Worksheet 2, page 11. This will help students collect the appropriate data and allow class data to be combined. You may need to make additional copies of Worksheet 2.
- Have students read the Chapter 1 Project Overview, pages 8–9. Review the project's rules and hand out the Chapter 1 Scoring Rubric, page 12, you will use for scoring students' work. Discuss with students what will be expected of them.
- Set deadlines for data analysis and presentations, and dates for interim stages of the project. Have students fill in the due dates on the Project Time Line.
- Make your own observations during the assignment. This will help you know whether students might be having problems because of cloud cover. Keep a class record. This helps students increase their data pool and can help those who were unable for some reason to record their own observations.

## ◆ Materials and Preparation

- Introduce describing the moon's position in the sky by locating objects in your room. You can identify anything on the wall by knowing which direction it is in and then knowing how high up the wall it is.
- To introduce compass direction, draw a compass on the blackboard. Explain how a compass works, and show how the compass rose is divided into eighths. Then take students outside, divide them into small groups, and have groups go to different parts of the school yard, with one compass per group. Have each group find landmarks at each of the eight compass directions.
- To introduce measurement of altitude in degrees, take students outside. Choose a tree to estimate altitude. Have students make a fist and hold it at arm's length. How many fists above the horizon is the top of the tree? One fist above the horizon is about 10°, two fists is about 20°, and so on. Remind them that objects directly overhead are at 90°, and all others are less than 90°.

## ◆ Check Your Progress—Section 1-1 Review

- Check students' direction maps of their observation sites, which they have made by completing Worksheet 1. Point out that if their landmarks are too near, the direction

**CHAPTER 1 PROJECT** *(continued)*

the landmarks are from the observer will change if the observer needs to move in order to get a better view of the moon.

- Check that students know how to complete the data sheet (Worksheet 2). Do students understand how to measure altitude?

## ◆ Check Your Progress—Section 1-2 Review

- Check that students are making observations and recording that data properly. Are they filling in all the required information? They will be starting to observe that the moon does not rise at the same time every day.

## ◆ Check Your Progress—Section 1-4 Review

- By this stage, students will have recorded enough observations to find patterns in the changes of the moon's shape. Have students examine two sketches of the moon made a few hours apart. Do students' sketches show that the orientation of the moon changes in the sky? The moon's crescent shape appears to rotate clockwise a bit during the day.

## ◆ Chapter Project Wrap-Up

- Have each student make a presentation based on their observations. The content and format of presentations may vary from student to student, depending on their data and interests. They may incorporate tables, maps, drawings, and graphs.
- Have students graph data from their observations. Assign or have students choose data to graph. Options include these: calendar date versus time of moon rise (phases of moon should be noted along with the date), compass direction versus time of observation in the course of one day, and compass direction versus frequency of sightings (this will show that the moon is only seen in certain parts of the sky, and never in others.
- In a class discussion, emphasize the following patterns:

  In the course of a day, the moon's position changes from eastern sky, through southern, to western (because of Earth's rotation).

  Moonrise gets progressively later in the day throughout the cycle.

  The moon will be seen mostly in the southern half of the sky, rising more easterly and setting more westerly.

  The lit portion of the moon starts on the right side and waxes until full. As it wanes, the right side becomes slowly dark.

## ◆ Extensions/Options

- Best results may be obtained if students share their observations daily, and class members add to their data lists the information from their classmates.
- Moon sightings made at school could be pointed out for all students to record.
- Moonrise and moonset times might be recorded from a newspaper weather page.
- At any location on Earth, an eclipse of the moon (partial or total) occurs at least once a year. If an eclipse will occur when your students are doing this project, be sure students are aware this is going to happen. If it happens early in the evening, the class might observe it together. Otherwise, students might get another adult to observe with them at home.

1

CHAPTER 1

# Where's the Moon?

This project requires you to keep records of when and in what direction you see the moon in the sky. You will make and record your observations for a number of weeks. Then you will look for patterns in your observations and explain the patterns by using what you know about the moon's orbit around Earth and Earth's rotation. To help keep your observations organized, you need to record the following information:

1. **Date** Record the calendar date (Example: November 10, 2001).
2. **Time of observation** For this project, you should try to make at least three observations per day. These observations should be two or three hours apart if possible.
3. **Moon visible** Try to make your observations from a place where you can see the sky in all directions. If you can see the moon, determine its direction, altitude, and appearance. Write this information on the data sheet as explained below. If you do not see the moon, the reason might be that the moon is not in the sky or that it is hidden behind clouds. If the sky is too cloudy to tell whether the moon is above the horizon, record that information in your data sheet.
4. **Position** The moon travels a path through the sky as if it were an object moving on a dome over Earth. Any position on the dome can be described by its direction from the observation location and its altitude above the horizon.

   Directions follow the compass headings. A compass fixes one direction, north (N), and then all directions along the circle of the horizon can be described.

   The circle of the compass is divided into four directions, north (N), south (S), east (E), and west (W). Four other headings that fall exactly between the four main directions are usually marked on compasses—northwest (NW), northeast (NE), southwest (SW), and southeast (SE). For your observations, record in which of these eight directions you see the moon. Worksheet 1 guides you in making a map of your moon observatory and marking the compass directions on it.

   Altitude normally measures how high an object is above the ground. For very distant objects, a more useful way to record position in the sky is to measure the altitude in term of degrees above the horizon. When an object is on the horizon, its altitude is zero degrees (0°). When an object is directly overhead, its altitude is 90°. You can easily estimate altitude in degrees. Face in the direction of the moon, make a fist and hold it straight out at arm's length, thumb side up. Count how many fists above the horizon the moon is. One fist above the horizon is about 10°, two fists are about 20°, and so on.

5. **Appearance** Make a sketch of the moon by shading in the circle. As you shade in the circle to show the moon's phase, try to sketch any slant or tilt of the bright part of the moon. When you are finished, the white part of your circle will look like the bright portion of the moon that you observed.



## ◆ Project Rules

- Record observations every day.
- Make observations more than once per day. Try to schedule times that will be easy to remember. For example you could look at the sky just before sunrise, just after sunset, and then right before you go to bed.
- Record all of the data listed in the table on Worksheet 2. Also use the worksheet to record any notes about your observations.
- After all observation data has been recorded, analyze the data. Make a class presentation about your observations. Your teacher will help you decide what should be included in your presentation. Your teacher will approve the idea for your presentation before you begin working on it.

## ◆ Project Hints

- The best site for viewing the moon will have the night sky visible in all directions.
- Try to find an observation site that is away from bright lights so that you can see the stars on a clear night. This will allow you to determine whether the sky is cloudy (no stars visible).
- Complete Worksheet 1 to identify landmarks so that you can determine each of the eight compass directions.

## ◆ Project Time Line

| Tasks | Due Date | Teacher's Initials |
|---|---|---|
| **1.** Direction Map (Worksheet 1) completed | ________ | |
| **2.** Moon observations begun | ________ | |
| **3.** Teacher check of data table | ________ | ________ |
| **4.** Observations completed | ________ | |
| **5.** Data analyzed | ________ | |
| **6.** Presentation made | ________ | |

Name ______________________ Date __________ Class ______________

CHAPTER 1 PROJECT WORKSHEET 1

# Make a Direction Map

1

1. Obtain or borrow a compass with a face that is at least 2 centimeters in diameter and that has marks for the eight directions (N, NE, E, SE, S, SW, W, NW).
2. During daylight hours, go to the site that you will use to record the moon observation data. On a blank sheet of paper, draw a large circle. Draw lines to divide the circle into halves, then quarters, and finally eighths. The center of the circle represents your position. Label your circle with the eight points of the compass.
3. Place the compass on a flat surface, such as a book. Make sure that the surface is level so that the compass needle is free to spin. Keep the compass away from metal objects. They will affect the compass readings.
4. Allow the compass needle to come to rest. One end of the needle (usually the end that is red or pointed) will be pointing north. Rotate the body of the compass until the needle lines up with the "N" on the face. Stand behind the needle and look in the "N" direction (the needle will be pointing in this direction). Identify a landmark, one far away if possible, and mark it on your drawn circle. Label it "north." If there is no landmark exactly north of your location, pick a close one and put it in the approximate location on your direction map.
5. Now stand on the opposite side of the compass and look in the "S" direction. The needle and the "N" of the compass need to be lined up again. The back of the compass needle will be pointing in south. Identify a distant landmark and mark it on your map. Do the same for all eight directions.
6. To check your direction map, stand at your observation location and throw an object, such as a Frisbee, in any direction. Once it lands, *use your map* to determine in what compass direction the object lies relative to where you are standing. Next use the compass to check the direction you obtained from your map (don't forget to align the needle and the "N"). If your map direction and compass direction do not agree, make the necessary corrections to your map.

Name ______________________ Date __________ Class ______________



# Moon Observation Data Sheet

| Date | Time | Moon Visible? (yes, no) | Direction (N, S, etc.) | Altitude (degrees) | Appearance (sketch) | Notes (for example, explain if not visible) |
|---|---|---|---|---|---|---|
| | | | | | | |
| | | | | | | |
| | | | | | | |
| | | | | | | |
| | | | | | | |
| | | | | | | |
| | | | | | | |
| | | | | | | |
| | | | | | | |
| | | | | | | |
| | | | | | | |
| | | | | | | |
| | | | | | | |
| | | | | | | |
| | | | | | | |
| | | | | | | |
| | | | | | | |

CHAPTER 1

# Where's the Moon?

In evaluating how well you complete the Chapter 1 Project, your teacher will judge your work in four categories. In each, a score of 4 is the best rating.

| | 4 | 3 | 2 | 1 |
|---|---|---|---|---|
| ***Moon Observations and Daily Record Keeping*** | Observations made daily. Data table complete and neat with accurate sketches and notation of cloudy skies. | Observations made almost daily. A few data are incomplete or inaccurate. No distinction between cloudy skies and clear skies when moon below horizon. | Most days have observations. Quite a few data are incomplete or inaccurate. No distinction between cloudy skies and clear skies when moon below horizon. | Many days have no observations. Many errors and omissions. Data table sloppy. |
| ***Graphs and Analysis*** | Student makes complete and correct analysis of data. Graph is correct and clear. | Student makes mostly complete and correct analysis of data. Graph is correct. | Student's analysis of data is mostly correct but is partially incomplete. Graph may have minor errors. | Data analysis is minimal or significantly incorrect. Graph has at least one significant error. |
| ***Concept Understanding and Presentation of Results*** | Data analysis and presentation are thorough and indicate full understanding of how and why the position and appearance of the moon change over time. | Data analysis and presentation are adequate and indicate good understanding of how and why the position and appearance of the moon change over time. | Data analysis and presentation are acceptable and indicate some understanding of how and why the position and appearance of the moon change over time. | Data analysis and presentation are minimal and indicate important misunderstanding of how and why the position and appearance of the moon change over time. |
| ***Communication and Class Participation*** | Communicates clearly and accurately. Participates fully in class discussion. | Communication is fairly clear and accurate. Class participation is adequate. | Communication mostly accurate but is limited. Participation is limited. | Communication is unclear or mostly inaccurate and class participation is minimal. |

# Earth in Space (pp. 14-23)

## Objectives

After completing the lesson, students will be able to

- explain what causes day and night;
- explain what causes the cycle of seasons on Earth.

## 1 Engage/Explore

- **Activating Prior Knowledge** Ask students to estimate what time the sun rises and sets and to describe how the number of hours of daylight each day change during the winter and summer.
- **Discover** Invite students to observe day and night using a globe and light bulb.

## 2 Facilitate

- **Building Inquiry Skills: Making Models** Help students use a lamp and globe to model the motion of Earth around the sun.
- **Sharpen Your Skills** Ask students to calculate what distance the Earth travels in a minute, an hours, a day, and a year.
- **Demonstration** Model Foucault's pendulum on a turntable and challenge students to explain how this models a pendulum at the North Pole.
- **Real-Life Learning** Have students examine the Yellow Pages to identify companies whose work is primarily seasonal.
- **Science and History** Challenge students to think about how the physical structures shown could serve as calendars.
- **Using the Visuals: Figure 2** Guide students to compare the area of Earth's surface covered by the sun's rays at the equator and the poles and how this affects the climates at those places.
- **Building Inquiry Skills: Comparing and Contrasting** Help students compare the sunlight at the equator and poles using a flashlight and large sheets of graph paper.
- **Math Toolbox** Direct students to calculate what fraction of a circle 90° is.
- **Exploring the Seasons** Ask students what causes it to be summer and winter in the northern hemisphere.
- **Demonstration** Demonstrate the changing seasons in the northern hemisphere using a globe and light bulb.
- **Inquiry Challenge** Challenge students to make models to test the hypothesis: *If Earth were not tilted, the length of the days would not change.*
- **Addressing Naive Conceptions** Ensure students know that the days are not longer in summer because of daylight-savings time.
- **Integrating Life Science** Encourage students to think about how the change in the length of days affects humans. Then have them brainstorm a list of animals or events that are unique to each season.

## 3 Assess

- **Section 1 Review** Check students' understanding of Earth in space by having them answer the section review questions.
- **Performance Assessment** Assess students' understanding of day, night, and seasons by having them choose an unfamiliar place on the globe and write descriptions of the amount of sunlight received there throughout the year.

## Lab

- **Skills Lab: Observing** Challenge students to use an Earth-sun model to make observations about factors that contribute to the seasons.

1

SECTION 1-1 SECTION SUMMARY

# Earth in Space

1

**Guide for Reading**

- What causes day and night?
- What causes the cycle of seasons on Earth?

The study of the moon, stars, and other objects in space is called **astronomy.** Ancient astronomers studied the movements of the sun and moon. They thought Earth was standing still and the sun and moon were moving. The sun and moon only seem to move because Earth is rotating on its axis. Earth's **axis** is an imaginary line that passes through Earth's center and the North and South poles. The spinning of Earth on its axis is called its **rotation. Earth's rotation on its axis causes day and night. It takes Earth about 24 hours to rotate once on its axis.**

The movement of one object around another object is called **revolution.** Earth completes one revolution around the sun once every year. Earth's path as it revolves around the sun is called its orbit. Earth's orbit is a slightly flattened circle.

Many cultures have tried to make a workable calendar. This is not easy because Earth takes about 365 1/4 days to circle the sun and 12 moon cycles make up fewer days than a year.

Sunlight hits Earth's surface most directly at the equator. Closer to the poles, sunlight hits Earth's surface at an angle. That is why it is warmer near the equator than near the poles.

**Earth has seasons because its axis is tilted as it moves around the sun.** Earth's axis is tilted at an angle of 23.5° from vertical. As Earth revolves around the sun, its axis is tilted away from the sun for part of the year and toward the sun for part of the year. When the Northern Hemisphere is tilted toward the sun, that half of the Earth has summer. At the same time, the Southern Hemisphere is tilted away from the sun and has winter. The hemisphere tilted toward the sun has more daylight hours than the hemisphere tilted away from the sun. The combination of direct rays and more hours of sunlight heats the surface more than at any other time of the year. In June, the north end of Earth's axis is tilted toward the sun.

**Latitude** is a measurement of distance from the equator, expressed in degrees north or south. On two days each year, the noon sun is overhead at either 23.5° north or south. Each of these days is known as a **solstice.** About June 21, the noon sun is directly overhead at 23.5° north latitude. Halfway between the solstices, neither hemisphere is tilted toward the sun. On those two days of the year, the noon sun is directly overhead at the equator. Each of these days is known as an **equinox,** which means "equal night," because the length of nighttime and daytime are about the same. The **vernal equinox,** or spring equinox, occurs around March 21, and marks the beginning of spring in the Northern Hemisphere. The **autumnal equinox** occurs around September 21. It marks the beginning of fall in the Northern Hemisphere.

Name ________________ Date ________ Class ________

SECTION 1-1 REVIEW AND REINFORCE

# Earth in Space

## ◆ Understanding Main Ideas

*Use the following figure to answer questions 1 through 3. Write your answers on the back of this page or on a separate sheet of paper.*

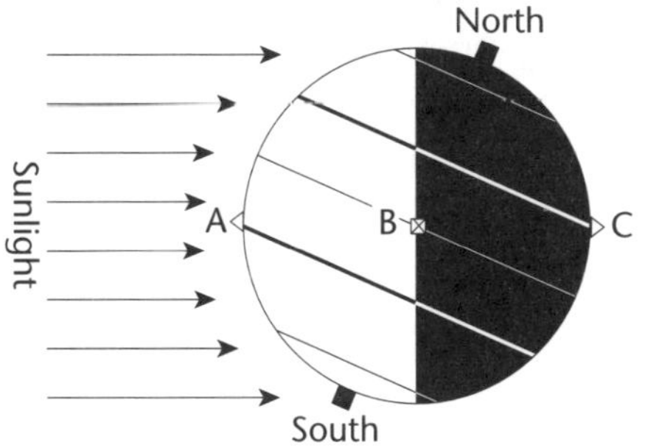


1. In the diagram, what season is it in North America?
2. Would a person at each of the points A, B, and C see the sun? If so, where would the sun be in the sky?
3. Which is a person standing at point B seeing, sunrise or the sunset? Explain.

## ◆ Building Vocabulary

*Match each term with its definition by writing the letter of the correct definition in the right column on the line beside the term in the left column.*

_____ 4. astronomy
_____ 5. axis
_____ 6. rotation
_____ 7. revolution
_____ 8. orbit
_____ 9. latitude
_____ 10. equinox
_____ 11. solstice
_____ 12. vernal equinox
_____ 13. autumnal equinox

**a.** The path of Earth as it revolves around the sun
**b.** Occurs in September and marks the beginning of fall in the Northern Hemisphere
**c.** Occurs in March and marks the beginning of spring in the Northern Hemisphere
**d.** The study of the moon, stars, and other objects in space
**e.** The sun is directly overhead at 23.5 degrees north or south at this time.
**f.** Movement of Earth around the sun
**g.** Movement of Earth around its axis
**h.** The sun is directly overhead at the equator at this time.
**i.** Line passing through Earth's center and poles
**j.** A measurement of distance from the equator, expressed in degrees north or south

Name ______________________________ Date __________ Class ________________

**SECTION 1-1** **ENRICH**

# Build a Simple Sundial

Long before the invention of clocks, people measured time with *sundials.* A simple sundial is a vertical pointer projecting from a flat surface. Marks on the surface show where the shadow cast by the tip of the pointer falls at different times of day as Earth's rotation causes the sun to move across the sky. You tell time by noting where the pointer's shadow falls on these marks. Of course, for the shadow to fall in the right places, the sundial has to stay in one place. Therefore, sundials are usually built on permanent pedestals or are parts of buildings. Also, the position of the sun in the sky changes through the seasons, so the path of the pointer's shadow changes over time.

To build a simple sundial, you can start with two pieces of stiff posterboard and masking tape.

- ◆ On the first piece of posterboard, draw the figure shown in Figure 1. Cut it out, then fold it along the dotted lines. This will become your pointer.
- ◆ On the second piece of posterboard, draw the figure shown in Figure 2, with the point of the angle close to a corner of the posterboard. Now, tape the pointer onto the second piece of posterboard so that its base lines up with the figure you have drawn. The finished sundial should look like Figure 3.
- ◆ Around midday, place the sundial on a stable, flat, sunlit surface such as an outdoor table top, so that the pointer's shadow falls towards the opposite corner of the posterboard. Tape down the sundial so it won't move. Every half hour, mark the location of the shadow of the pointer's tip and label it with the time. Continue this process throughout the afternoon and again the next morning. This is how you "set your clock," when your clock is a sundial.

Figure 1

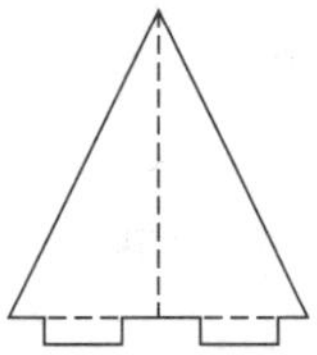

Base: 20 cm.
Height: 20 cm.

Figure 2

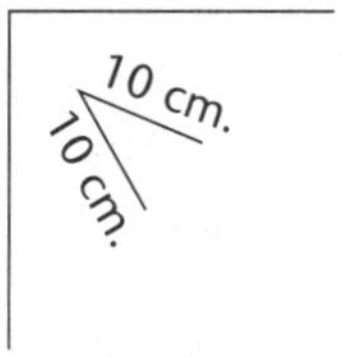


Figure 3

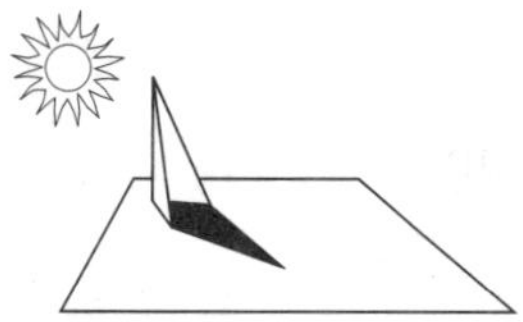

*Answer the questions below on the back of this sheet or on a separate sheet of paper.*

**1.** At some moment when you don't know what time it is, look at your sundial and read the time. Compare your sundial time with the time from a clock. How accurate was your sundial?

**2.** Do you think a "moondial" could be made to tell time at night? Explain.

**3.** Name two practical disadvantages of sundials compared with clocks.

# Phases, Eclipses, and Tides (pp. 24-34)

## Objectives

After completing the lesson, students will be able to

- explain what causes the phases of the moon;
- compare the causes of solar and lunar eclipses;
- explain what causes the tides.

## 1 Engage/Explore

- **Activating Prior Knowledge** Ask students to describe observations they have made of the moon.
- **Discover** Challenge students to infer if from the point of view of someone on Earth, does the moon seem to rotate.

## 2 Facilitate

- **Including All Students** Help students understand why the flags in Figure 5 indicate that the moon is rotating.
- **Social Studies Connection** Ask how the harvest moon got its name.
- **Exploring Phases of the Moon** Guide students in exploring moon phases.
- **Inquiry Challenge** Invite students to predict the days between the new moon, first quarter, full moon, and third quarter.
- **Building Inquiry Skills: Comparing and Contrasting** Have students list and define three ways *phase* is used in science.
- **Using the Visuals: Figure 6** Ask students what phase the moon must be in for a solar and lunar eclipse to occur.
- **Language Arts Connection** Have students read *A Connecticut Yankee in King Arthur's Court* and explain how lack of scientific explanations might encourage superstitious beliefs.
- **Cultural Diversity** Discuss where some of the terms in this section come from.
- **Sharpen Your Skills** Direct students to draw a scale model of a solar eclipse.
- **Using the Visuals: Figure 7** Ask students if people in the moon's penumbra experience a total or a partial eclipse.
- **Using the Visuals: Figure 8** Have students interpret why we don't have a lunar eclipse during every full moon.
- **Building Inquiry Skills: Comparing and Contrasting** Challenge students to compare and contrast the umbra and penumbra of the moon during a total and partial lunar eclipse.

## Lab

- **Skills Lab: Making Models** Challenge students to use a model of the Earth-moon-sun system to explore moon phases.
- **Building Inquiry Skills: Making Models** Direct students to model the moon's pull on Earth with a balloon.
- **Building Inquiry Skills: Predicting** Encourage students to predict times of high and low tides from tide tables.
- **Using the Visuals: Figure 10** Ask students how many low and high tides one particular spot has in one day.
- **Inquiry Challenge** Challenge students to model the combined force of the sun's and moon's gravity on Earth.
- **Integrating Life Science** Point out that organisms living in the intertidal zone spend part of each day submerged in water, and part of the day exposed to sunlight and air.

## 3 Assess

- **Section 2 Review** Check students' understanding of section content by having them answer the section review questions.
- **Performance Assessment** Assess students' understanding of phases, eclipses, and tides by having them demonstrate the positions of Earth, the sun, and the moon either during a solar eclipse, a lunar eclipse, or during high and low tides.

1

Name ______________________ Date __________ Class __________

SECTION 1-2 SECTION SUMMARY

# Phases, Eclipses, and Tides

1

**Guide for Reading**

- What causes the phases of the moon?
- What causes solar and lunar eclipses?
- What causes the tides?

As the moon moves, the positions of the moon, Earth, and the sun change in relation to each other. **The positions of the moon, Earth, and the sun cause the phases of the moon, eclipses, and tides.**

The moon revolves around Earth about once every 27.3 days. It also rotates on its own axis about once every 27.3 days. The same side of the moon always faces Earth. The different shapes of the moon you see from Earth are called **phases.** Phases are caused by changes in the relative position of the moon, Earth, and sun. **The phase of the moon you see depends on how much of the sunlit side of the moon faces Earth.**

**When the moon's shadow hits Earth or Earth's shadow hits the moon, an eclipse occurs.** An **eclipse** occurs when an object in space comes between the sun and a third object, and casts a shadow on that object. There are two types of eclipses: solar and lunar.

A **solar eclipse** occurs when the moon passes between Earth and the sun, blocking the sunlight from reaching Earth. The moon's shadow then hits Earth. The darkest part of the moon's shadow is called the **umbra.** From any part of the umbra, the moon completely blocks light from the sun. Only people in the umbra see a total solar eclipse. Another part of the shadow is less dark and larger than the umbra. It is called the **penumbra.** From within the penumbra, people see a partial eclipse, because part of the sun is still visible.

A **lunar eclipse** occurs at a full moon when Earth is directly between the moon and the sun. During a lunar eclipse, Earth's shadow falls on the moon. Earth's shadow also has an umbra and a penumbra. When the moon is completely within Earth's umbra you see a total lunar eclipse. A partial lunar eclipse happens when the moon moves into Earth's penumbra.

**Tides** are the rise and fall of the ocean's water every 12.5 hours or so. The force of **gravity** pulls the moon and Earth toward each other. **Tides occur mainly because of differences in how much the moon pulls on different parts of Earth.** As Earth rotates, the moon's gravity pulls water toward the point on Earth's surface closest to the moon. The moon pulls least on the side of Earth farthest away. Two tides occur each day because of this difference in the pull of the moon's gravity.

Twice a month, the moon, Earth, and the sun are in a straight line. The combined forces of the gravity of the sun and moon produce an especially high tide—called a **spring tide**—and an especially low tide. Also twice a month, the pull of gravity of the sun and moon are at right angles to each other. At those times the high tide is lower than usual, and is called a **neap tide.** The low tides then are higher than usual.

Name ______________________ Date ______ Class ______________

SECTION 1-2 REVIEW AND REINFORCE

# Phases, Eclipses, and Tides

## ◆ Understanding Main Ideas

*Use the following figure to answer questions 1 and 2. Write your answers on the back of this page or on a separate sheet of paper.*

1. What phases of the moon would someone on Earth see when the moon is at positions A through F?
2. What kind of tide will occur when the moon is at positions A, C, D, and F?

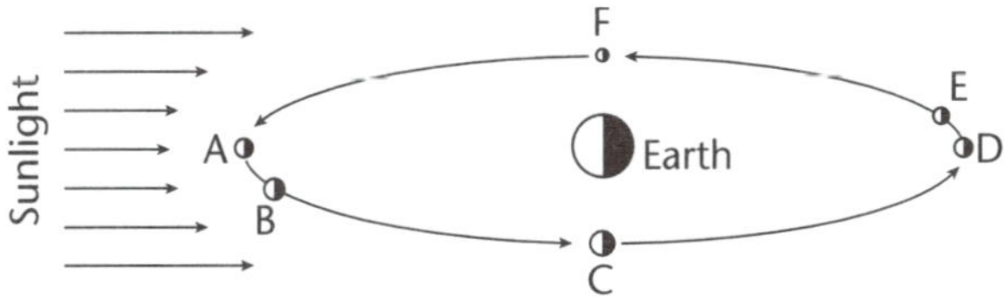


## ◆ Building Vocabulary

*From the list below, choose the term that best completes each sentence.*

| | | | | |
|---|---|---|---|---|
| phase | tides | gravity | lunar | penumbra |
| eclipse | umbra | spring | solar | neap |

3. A(n) ______________ tide occurs when the sun is at right angles to the line between Earth and the moon.
4. A(n) ______________ occurs when the moon's shadow hits Earth or Earth's shadow hits the moon.
5. A person standing in the moon's ______________ would see a partial solar eclipse.
6. Differences in the moon's pull on different parts of Earth cause ______________.
7. A person standing in the moon's ______________ would see a total solar eclipse.
8. The ______________ of the moon you see depend on how much of the sunlit side of the moon faces Earth.
9. A(n) ______________ tide occurs when the sun, moon, and Earth line up.
10. A(n) ______________ eclipse occurs at a full moon when Earth is directly between the moon and the sun.
11. A(n) ______________ eclipse occurs when the moon passes between Earth and the sun.
12. The force of ______________ pulls the moon and Earth toward each other.

# Lunar Calendars

1 It is easy to see the changes in the appearance of the moon, so early peoples frequently based their calendars on the cycles of the moon. The word "month" comes from the same word as "moon." However, the months of our solar calendar are no longer lunar months, because the Western calendar has been adjusted to become a solar calendar. Since the moon takes about 29.5 days to complete a cycle of lunar phases, 12 lunar months is 11 days short of a solar year. In spite of this difference, many people still organize their lives by lunar calendars. Two of the world's great religions, Judaism and Islam, keep lunar calendars.

Most Muslims recognize a 12 lunar month Islamic religious calendar for their private lives, even if they keep a different civil calendar. Originally, each lunar month began when the new moon was first sighted in the evening sky at sunset. Recently, however, many Muslims worldwide recognize official timekeepers in Cairo, Egypt, who announce the beginning of each month.

Unlike many other calendars, no attempt is made to keep the Islamic religious calendar synchronized with a solar calendar. So the holy month of Ramadan, a time of prayer and fasting among all Muslims, continually moves through the seasons, occurring about 11 days earlier each solar year. One year, Ramadan might be in the winter, and 16 solar years later, it would be in the summer.

The Jewish calendar has 12 or 13 lunar months. The standard year is 12 months, but an extra (or *intercalary*) month is inserted seven times during a 19-year cycle to keep the religious calendar closely aligned with the solar year. Thus, Jewish religious holidays always occur in the same season, although the exact dates vary. For example, the Jewish holiday Passover is always in spring.

1. How many days are in a lunar "year" of 12 lunar months? How many days are in a solar year?
2. Not counting leap years, how many days are in 19 solar years? How many days are in 19 lunar calendar years with an extra month added in seven times? How closely does the Jewish calendar coincide with the solar calendar every 19 years?
3. Why might it be important for farmers and hunters to keep their calendar aligned with the seasons?

*Integrating Technology*

# Rockets and Satellites (pp. 35-38)

## Objectives

After completing the lesson, students will be able to

- explain how rockets travel in space;
- describe what satellites and space stations are used for.

## 1 Engage/Explore

- **Activating Prior Knowledge** Invite volunteers to describe what happens when a rocket is launched. Ask them to consider the motion of the rocket in their responses.
- **Discover** Encourage students to model a rocket with a balloon. Direct them to observe the direction the air rushes out and the direction the balloon goes.

## 2 Facilitate

- **Using the Visuals: Figure 13** Ask students what action and reaction is shown in the diagram.
- **Demonstration** Demonstrate a model of a multistage rocket using two balloons attached to fishing line run across the room.
- **Try This** Challenge students to build a rocket using a fizzing antacid tablet and water in a film canister to launch the rocket. Direct students to observe what action happened inside the film canister and what was the reaction of the rocket.
- **Inquiry Challenge** Encourage students to use physical activity to model a geosynchronous orbit and a low Earth orbit.
- **Addressing Naive Conceptions** Help students visualize the low-Earth orbit the shuttle was meant for by using a paper ruler perpendicular to a globe to show the highest altitude of the orbiting shuttle.

## 3 Assess

- **Section 3 Review** Check students' understanding of rockets and satellites by having them answer the section review questions.
- **Performance Assessment** Assess students' understanding of rockets and satellites by having them create a poster showing how a space station or satellite is launched into orbit and then used for practical applications on Earth.

1

Name ______________________ Date __________ Class ______________

SECTION 1-3 SECTION SUMMARY

# Rockets and Satellites

**Guide for Reading**

- How do rockets travel in space?
- What are satellites and space stations used for?

No space mission could be accomplished without rockets. **A rocket moves forward when gases expelled from the rear of the rocket push it in the opposite direction.**

Early rockets used gunpowder for fuel. Gunpowder burns quickly and explosively. A rocket used to travel out of Earth's atmosphere needs a fuel that burns more slowly and continuously. Dr. Robert H. Goddard experimented with liquid fuels and found that liquid fuels can provide continuous power.

Konstantin Tsiolkovsky came up with the idea of multistage rockets. As each stage, or section, of a rocket uses up its fuel, the empty fuel container drops off. Then the next stage ignites. The development of multistage rockets made it possible to send rockets to the moon and farther into space.

A **satellite** is any natural or artificial object that revolves around an object in space. The first artificial satellite to orbit Earth was *Sputnik 1.* The Soviet Union launched it in 1957. In 1961 a Soviet cosmonaut named Yuri Gagarin orbited Earth, becoming the first person in space. Since then, many artificial satellites have been sent into orbit by rockets or shuttles.

**Satellites and space stations are used for communications, navigation, collecting weather data, and research.** Some satellites move in **geosynchronous orbits,** which means that they revolve around Earth at the same rate that Earth rotates. Geosynchronous satellites seem to hover over one point on Earth. They are used to broadcast and relay television signals and map weather patterns.

A space station is a large satellite in which people can live for long periods of time. Sixteen countries, including the United States and Russia, are now cooperating to build the International Space Station.

The rockets that carried astronauts to the moon were very expensive and could not be reused. Since that time, the National Aeronautics and Space Administration (NASA) has developed the reusable space shuttle. They are called shuttles because they can go back and forth, or shuttle, between Earth and space. Since 1981, space shuttles have been the main way the United States launches astronauts and equipment into space.

Name ______________________ Date __________ Class __________

SECTION 1-3 REVIEW AND REINFORCE

# Rockets and Satellites

## ◆ Understanding Main Ideas

*Answer the following questions in the space below.*

**1.** What causes a rocket to move forward?

**2.** What is the difference between a multistage rocket and a single-stage rocket?

**3.** Name at least three uses of satellites and space stations.

**4.** What is the difference between space shuttles and other rockets?

## ◆ Building Vocabulary

*Write a brief description of each of the following.*

**5.** geosynchronous orbit

**6.** satellite

Name ____________________ Date __________ Class ____________



1

# Geosynchronous Orbits and Polar Orbits

When a satellite is in a geosynchronous orbit, its speed is such that it seems to hang over one spot on Earth. All satellites in geosynchronous orbit are about 36,000 km above Earth's surface. But a satellite's speed and altitude aren't the only requirements to be in a geosynchronous orbit—it must also be in orbit over the equator.

Another kind of orbit is called a *polar* orbit. This is an orbit that passes over both the North Pole and the South Pole.

On the map below, draw a line to represent the path a satellite in a low polar orbit might take as it passes over the United States. An easy way to do this is to start with your pencil on the **X** (it would be a place in Canada) and begin to move it straight south (down) to represent the direction of movement of the satellite. At the same time, move the paper to the east (right), to represent the motion of Earth. Your result will be a line moving from northeast to southwest. You can see that it is the combination of the satellite's motion and Earth's rotation that determines the path a satellite takes over Earth.

*Answer the following questions on a separate sheet of paper.*

1. A satellite in a polar orbit traces a curved path over the surface of Earth. Explain why.
2. Why doesn't a geosynchronous satellite trace a path over the surface of Earth?
3. Imagine that the satellite in polar orbit whose path you drew above takes two hours to complete one orbit. After two hours, the Earth will have rotated two hours to the east and the satellite will be passing over the place represented by the **Y** on the map. About where will it be in more two hours?
4. How often would the satellite in question 3 pass over the same point on Earth?
5. Television signals can't pass through Earth. If you wanted to launch a satellite that would be used to transmit television images from the United States to Europe 24 hours a day, would you put it into a geosynchronous orbit or a polar orbit? Explain.
6. If you wanted to launch a satellite that would be used to take photographs of the entire surface of the Earth, what kind of orbit would you put it in? Explain.

# Earth's Moon (pp. 39-44)

## Objectives

After completing the lesson, students will be able to

- describe features of the moon that can be seen with a telescope;
- explain how the Apollo landings helped scientists learn about the moon.

## 1 Engage/Explore

- **Activating Prior Knowledge** Engage student's interest by encouraging them to imagine a rock as large as a building falling through Earth's atmosphere and landing in an open desert. Ask them what they think would happen.
- **Discover** Allow students to experiment creating different sized craters by dropping marbles of different masses at various heights.

## 2 Facilitate

- **Using the Visuals: Figure 19** Have students organize the information in the figure into a flowchart. According to the theory, ask them what the moon is made of.
- **Building Inquiry Skills: Interpreting Data** After having students practice the indirect method of measuring the classroom clock, elicit explanations as to how scientists could use this method to find the diameter of the moon.
- **Including All Students** Help students who are mastering English to identify the features in Figure 20 using the text.
- **Visual Arts Connection** Ask students why they think that Galileo's observations were of such great historical importance.
- **Building Inquiry Skills: Applying Concepts** Have students brainstorm a list of factors that scientists need to consider when landing a human on the moon. Ask them what five items they would bring with them to the moon.
- **Social Studies Connection** While watching or listening to the first moon landing, invite students to describe how they would have felt during the tense last moments of the landing.
- **Using the Visuals: Figure 22** Ask students to describe challenges they think NASA scientists had to consider when creating a vehicle that could explore the moon's terrain.
- **Building Inquiry Skills: Developing a Hypothesis** Challenge students to imagine how the rocks on Earth could be used to test the theory that the moon was formed when an object collided with Earth.
- **Sharpen Your Skills** Ask students if they had to war a spacesuit that weighed as much as they did, what would be their total weight on the moon.

## 3 Assess

- **Section 4 Review** Check students' understanding of Earth's moon by having them answer the section review questions.
- **Performance Assessment** Assess students' understanding of Earth's moon by having them make a diagram of the moon using shadows to indicate the presence of craters and other features.

# Earth's Moon

1

**Guide for Reading**

- **What features of the moon can be seen with a telescope?**
- **How did the Apollo landings help scientists learn about the moon?**

The moon is 3,476 kilometers in diameter—about one fourth the diameter of Earth. However, the moon has only one-eightieth as much mass as Earth. Scientists do not know for sure how the moon was formed. The theory that best fits the evidence is called the collision theory. It says that about 4.5 billion years ago, an object about as large as Mars hit Earth. Material from the object and from Earth's outer layers was thrown into orbit around Earth. Eventually, that material formed the moon.

For thousands of years, people could see shapes on the surface of the moon, but didn't know what caused them. About 400 years ago, Galileo made his own **telescope** by putting two lenses in a wooden tube. He looked through his telescope at the moon, and he saw the moon in more detail than anyone had ever seen before. **Features on the moon's surface include craters, highlands, and maria.**

Galileo saw that much of the moon's surface is covered with round pits called **craters.** Galileo also saw dark, flat parts of the moon's surface which he called **maria,** the Latin word for "seas." The singular of maria is *mare.* Scientists now know that the maria are areas that were flooded by molten rock billions of years ago. The moon's craters were formed by impacts by rocks from space.

In 1961, President John F. Kennedy announced that America would send people to the moon. *Surveyor* was the first spacecraft to land on the moon. *Lunar Orbiters* then photographed the moon's surface, so scientists could find a safe spot for humans to land. On July 20, 1969, the *Eagle,* the Lunar Module of *Apollo 11,* carrying Neil Armstrong and Buzz Aldrin, landed in a flat region called the Sea of Tranquillity. Neil Armstrong was the first person to walk on the moon. Later, other *Apollo* missions also went to the moon.

Much of what scientists have learned about the moon came from detailed study of the moon rocks gathered by astronauts. Almost all the rocks were formed from the cooling of molten material. Some rocks showed they had been broken apart by impacts and then fused back together. Seismometers placed on the moon by the astronauts detected extremely weak moonquakes. Another instrument measured the amount of heat flowing from the moon's interior. It showed that the moon has cooled almost completely since it was formed. The far side of the moon is rougher than the near side and has few maria.

# Earth's Moon

## ◆ Understanding Main Ideas

*Answer the following questions on the back of this page or on a separate sheet of paper.*

1. How are the size and mass of the moon different from that of the Earth?
2. How do astronomers think the moon was formed?
3. Who was the first person to observe the moon through a telescope? What features of the moon did he identify?
4. Identify the major events in moon exploration marked on the time line below.

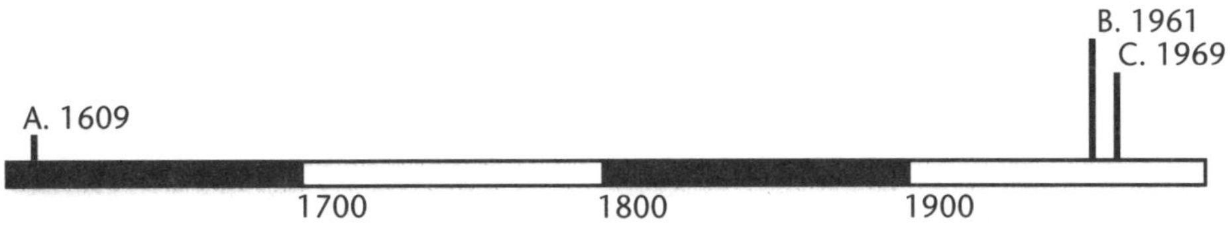


5. What new information about the moon did scientists learn from the Apollo mission?

## ◆ Building Vocabulary

*Answer the following questions in the space provided.*

6. How did Galileo make a telescope?

7. What are moon craters? How were they formed?

8. What are maria? How were they formed?

Name ______________________ Date __________ Class ______________

SECTION 1-4 ENRICH

# The Evolution of Lunar Landscapes

You already know that the surface of the moon has highlands and maria. Scientists think that the highlands of the moon could have formed a billion years before the maria. Remember, maria formed from molten rock that flowed across the moon's surface.

One way that astronomers estimate the age of a region on the moon is by counting its craters. If meteors are equally likely to hit any spot on the moon, older regions should have more craters per square kilometer because more meteors would have had a chance to strike them. This simple rule, along with a few others, enables astronomers to learn the order in which features on the moon were formed. Use it in the following activity.

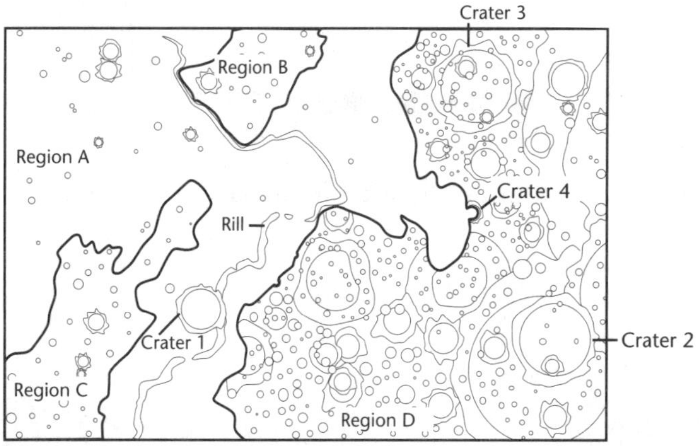


The Moon's Surface

*Answer the following questions on the back of this sheet or on a separate sheet of paper.*

1. In the map above, the heavy black lines mark the boundaries of four basic regions: A, B, C, and D. The circular features are craters. Based on the number of craters per unit of area, which region is probably the oldest? Which is probably the youngest?
2. Several small craters lie in the floor and on the walls of crater 2. Which is older, crater 2 or the small craters inside it?
3. Craters 2 and 3 have about the same area. Which one is probably older? Why?
4. Region A is a mare. The others are highlands. Notice how the material of region A fills crater 4. Which formed first, region A or crater 4? Why?
5. The long structures in region A are *rills.* A rill is a kind of valley formed when an underground tube collapses. Is the labeled rill older or younger than region A? List the following in order of age, from oldest to youngest: crater 1, crater 4, the labeled rill, and the floor of region A.

*Observing*

# Reasons for the Seasons

In this lab, you will use an Earth-sun model to make observations about factors that contribute to the seasons.

1

## ◆ Problem

What effect does the tilt of Earth's axis have on the heat and light received by Earth as it revolves around the sun?

## ◆ Materials (per pair of students)

books flashlight paper
pencil protractor toothpick
acetate sheet with thick grid lines drawn on it
plastic foam ball marked with poles and equator

## ◆ Procedure

1. Make a pile of books about 15 cm high. Dim the room lights.
2. Tape the acetate sheet to the head of the flashlight. Place the flashlight on the pile of books.
3. Carefully push a pencil into the South Pole of the plastic foam ball, which represents Earth.
4. Use the protractor to measure a 23.5° tilt of the axis of your Earth away from your "flashlight sun," as shown in the first diagram. This represents winter.
5. Hold the pencil so that Earth is steady at this 23.5° angle and about 15 cm from the flashlight head. Turn the flashlight on.
6. The squares on the acetate should show up on Earth. Move the ball closer if necessary or dim the room lights more. Observe the shape of the squares at the equator and at the poles. Record these in the Data Table on the next page.
7. Carefully stick the toothpick straight into Earth about halfway between the equator and the North Pole. Observe and record the length of the shadow.
8. Without changing the tilt, turn the pencil to rotate Earth once on its axis. Observe and record how the shadow of the toothpick changes.
9. Tilt Earth 23.5° toward the flashlight, as shown in the second diagram. This is summer. Observe and record the shape of the squares at the equator and at the poles. Observe how the toothpick's shadow changes.
10. Rotate Earth and note the shadow pattern.

Name ______________________ Date __________ Class ______________

## SKILLS LAB 1 *(continued)*

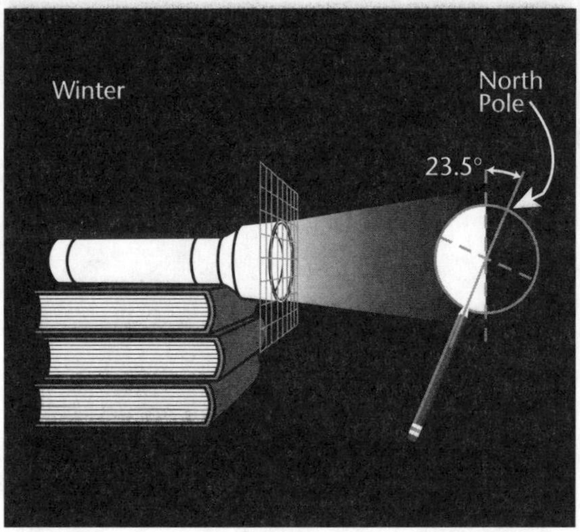


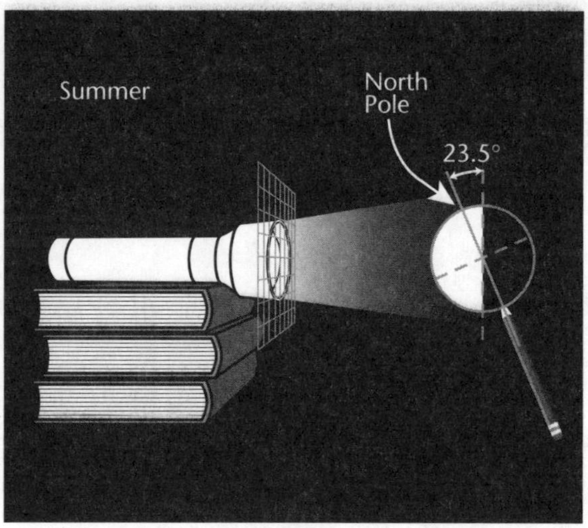


### ◆ Data Table

| | |
|---|---|
| **Winter** | |
| Shape of squares at equator | |
| Shape of squares at poles | |
| Length of shadow of toothpick | |
| Length of shadow of toothpick after rotation | |
| **Summer** | |
| Shape of squares at equator | |
| Shape of squares at poles | |
| Length of shadow of toothpick | |
| Length of shadow of toothpick after rotation | |

**SKILLS LAB 1** *(continued)*

## ◆ Analyze and Conclude

*Write your answers on the back of this sheet or on a separate sheet of paper.*

1. When it is winter in the Northern Hemisphere, which areas on Earth get the most concentrated amount of light? Which areas get the most concentrated light when it is summer in the Northern Hemisphere?
2. Compare your observations of how the light hits the area halfway between the equator and the North Pole during winter (Step 6) and during summer (Step 9).
3. If the squares projected on the ball from the acetate become larger, what can you conclude about the amount of heat distributed in each square?
4. According to your observations, which areas on Earth are consistently coolest? Which areas are consistently warmest? Why?
5. What time of year will the toothpick's shadow be longest? When will the shadow be shortest?
6. How are the amounts of heat and light received in a square related to the angle of the sun's rays?
7. **Think About It** How can you use your observations of an Earth-sun model to explain what causes the seasons?

## ◆ More to Explore

You can measure how directly light from the sun hits Earth's surface by making a shadow stick. You need a stick or pole about 1 m long. With the help of your teacher, push the stick partway into the ground where it will not be disturbed. Make sure the stick stays vertical. At noon on the first day of every month, measure the length of the stick's shadow. The shorter the shadow, the more directly the sun's rays are hitting Earth. At what time of the year are the shadows longest? Shortest? How do your observations help explain the seasons?

Name ______________________ Date __________ Class ______________



*Making Models*

# A "Moonth" of Phases

1

In this lab, you will use a model of the Earth-moon-sun system to explore how the phases of the moon occur.

## ◆ Problem

What causes the phases of the moon?

## ◆ Materials

floor lamp with 150-watt bulb
extension cord
pencils
plastic foam balls

## ◆ Procedure

*Review the safety guidelines in Appendix A.*

1. Place a lamp in the center of the room. Remove the lampshade.
2. Close the doors and shades to darken the room, and switch on the lamp.
3. Carefully stick the point of a pencil into the plastic foam ball so that the pencil can be used as a "handle".
4. Use the data table to record your data.
5. Have your partner hold the plastic foam ball at arm's length in front and slightly above his or her head so that the ball is between him or her and the lamp. (**CAUTION:** *Do not look directly at the bulb.*)
6. The ball should be about 1 to 1.5 m away from the lamp. Adjust the distance between the ball and the lamp so that the light shines brightly on the ball.
7. Stand directly behind your partner and observe what part of the ball facing you is lit by the lamp. If light is visible on the ball, draw the shape of the lighted part of the ball on the Data Sheet provided on the next page.
8. Have your partner turn 45° to the left while keeping the ball in front and at arm's length.
9. Repeat Step 7. Be sure you are standing directly behind your partner.
10. Repeat Steps 8 and 9 six more times until your partner is facing the lamp again. See the diagram for the 8 positions.
11. Change places and repeat Steps 4–10.

## ◆ Data Sheet

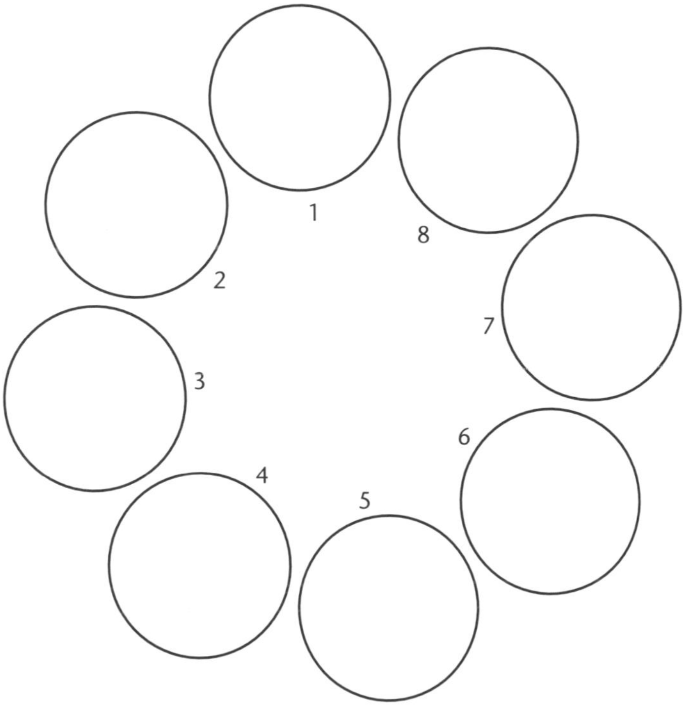


## ◆ Analyze and Conclude

*Write your answers on the back of this sheet or on a separate sheet of paper.*

1. In your model, what represents Earth? The sun? The moon?
2. Refer back to your Data Sheet. How much of the lighted part of the ball did you see when facing the lamp?
3. Label your Data Sheet with the name of the phases of the moon. Which drawing represents a full moon? A new moon? Which represents a waxing crescent? A waning crescent?
4. How much of the lighted part of the ball did you see after each turn?
5. Whether you could see it or not, how much of the ball's surface was always lit by the lamp? Was the darkness of the new moon caused by an eclipse? Explain your answer.
6. **Think It Over** How did making a model help you understand the phases of the moon? What are some disadvantages of using models? What is another way to make a model to represent the moon's phases?

## ◆ More to Explore

Design a model to show a lunar eclipse and a solar eclipse. What objects would you use for Earth, the sun, and the moon? Use the model to demonstrate why there isn't an eclipse every full moon and new moon.

CHAPTER 2

# Model of the Solar System



The following steps will walk you through the Chapter 2 Project. Use the hints as you guide your students through planning, model building, and presentations.

- Before introducing the project, bring a model, such as a miniature Statue of Liberty, into the classroom to show to the students. Talk about size and scaling. Measure features of the model, such as the statue's nose, and compare these measurements to the actual object.
- Have students read the Chapter 2 Project Overview, pages 36–37. Review the project's rules and hand out the Chapter 2 Scoring Rubric, page 40, that you will use to score students' work. Discuss with students what is expected of them.
- Set a deadline for the project presentation and interim dates for the Check Your Progress at the end of Section 1, Section 3, and Section 4. Encourage students to copy the dates in the Project Time Line.
- Distribute copies of Chapter 2 Project Worksheet 1, page 38. This worksheet should help them to think about scaling. Depending on the students' level of mathematical skill, you may decide to let them work on this worksheet in groups. This will give you more of an opportunity to assist them. Chapter 2 Project Worksheet 2, page 39, will help students get started with their projects as they begin to plan out their models. Distribute this worksheet once you are sure that all of the students understand the concepts of size and scaling. Remind students to refer to the hints in the Chapter 2 Project Overview as they plan and carry out the project.

## ◆ Materials and Preparation

- Consider organizing the class in groups to do the project. If space and time are limited, you may want to have all of the students work together on one large model. Consider allowing them to create their models outside where space is less limited.
- Meter sticks or metric rulers, tape measures, and calculators will help students with their model building. Objects such as beach balls, basketballs, baseballs, tennis balls, fruit (grapefruit, oranges), marbles, dried beans, ball bearings, and sand would make good models of planets, depending on their diameter and the scaling used by the students. For the model in which they demonstrate the distance between planets, chalk drawings on the chalkboard or a sidewalk outside will allow space for students to construct their models. City maps or a neighborhood map you create will also be useful.
- Additional materials that may assist students in thinking about size and scaling are available from such companies as the Astronomical Society of the Pacific, 390 Ashton Avenue, San Francisco, CA 94112, USA. The video *Powers of Ten* by Charles and Ray Eames may help students with their mathematical skills. It is available from the same location. A solar system chart that provides relative sizes, distances, orbits, and data about the planets and their place in the galaxy may help them to think about the size of the solar system.

## ◆ Check Your Progress—Section 1 Review

- A major difficulty of this project will be the mathematics of manipulating large numbers. Monitor students carefully to make sure they are calculating correctly. Review with students how to make conversions from kilometers to centimeters. For this part of the project, appropriately scaled distances can be found by dividing actual distances by 1,000,000,000,000 and expressing the final number in centimeters.

**CHAPTER 2 PROJECT** *(continued)*

- Make sure that students understand the concept of scaling. You could ask students to draw a model of the classroom where 1 centimeter = 1 foot. Explain that the ratio 1 cm : 1 ft represents that scale. Have students diagram their plan for demonstrating the distances between the planets.
- Review students' plans before allowing them to begin on this project. If they are planning on using a common school area, such as a hallway or sidewalk, make sure that they have obtained permission to do so.
- Check students' completed data tables at the top of Worksheet 1. Make sure that they have selected an appropriate scale, such as those suggested previously. Check their mathematical calculations to ensure that students understand how to proportionally scale the distances between the planets.

## ◆ Check Your Progress—Section 3 Review

- Before having students complete the data tables at the bottom of Worksheet 1, you may want to review the SI and have students practice converting between meters, centimeters, and millimeters.
- Check students' completed data tables at the bottom of Worksheet 1. Make sure that they have selected a scale for which they can chose object of the correct size. One useful scale can be found by dividing actual diameters by 1,000,000,000—the results are in centimeters. Check mathematical calculations to ensure that students understand how to proportionally scale the diameters of the planets.
- Check students' model plans. Be sure students have thought about objects that they can use as model planets.

## ◆ Check Your Progress—Section 4 Review

- Check students' completed data tables on Worksheet 2. Make sure that they have considered both the diameters of the planets and the distances between the planets.
- Make sure that students have considered the difficulties in creating one model that illustrates both size and distance. Have they thought about how they will illustrate this? Have they considered whether it is possible to do within the limits of the classroom?

## ◆ Chapter Project Wrap-Up

- As students demonstrate their models, ask their classmates to take brief notes, writing down the major differences between models. Did everyone use the same scales? Did some scales work better than others?
- After all the presentations have been made, have students discuss the different models. Have them decide which scales worked best for the project and what problems they encountered in building their models. Would they do anything differently the next time they build such models? Also discuss the vastness of space. Students should now have a better understanding of the size of the solar system and how much of it is empty space.

## ◆ Extensions/Options

- Consider building a large model of the solar system as a class activity. Model the solar system's objects with more detail. Individual students or groups of students could be responsible for modeling particular planets, moons, asteroids, and comets. This could be done as a chalk drawing outside on the school sidewalk, as a larger papier mâché model, or with objects of various sizes in the school yard.
- At the end of the project, take your students to a local planetarium where they can learn more about the planets in our solar system, as well as other astronomical topics.

Name ______________________ Date __________ Class ______________

CHAPTER 2 PROJECT OVERVIEW

CHAPTER 2

# Model of the Solar System

In this project, you make models of our solar system in three different ways. In one model, you will represent the distance of each planet from the sun. In a second model, you will illustrate the relative size of the sun and planets. Finally, you will combine the concepts shown in the first two models. At the end of the project, you will present your models to the class.

2

## ◆ Project Rules

- The first model will show the relative distance from each planet to the sun.
- The second model will show the relative sizes of the planets.
- The third model will show both the planets' relative distances from the sun and their diameters.
- You must build your models to scale. The scale should be clearly posted as part of each model.
- Have your teacher approve your research and plans before you begin each model.

## ◆ Suggested Materials

- Meter sticks or metric rulers, tape measures, and calculators will be needed for model building.
- For the model in which you demonstrate the distances between planets, drawings on the chalkboard, on cash register tape attached to a wall, or on a sidewalk will give you space to construct the models. You may also want to represent the positions of the planets using a community map. Mark a central landmark or building as the sun.
- Objects such as beach balls, basketballs, baseballs, tennis balls, fruit (grapefruit, oranges), marbles, dried beans, ball bearings, and grains of sand would make good models of planets, depending on their diameter and the scale used.

Name ______________________ Date __________ Class ______________



## ◆ Project Hints

- Use the worksheets to help you start thinking about possible scales to use for your models.
- When working with the distances between the planets, it will be very difficult to fit all of the planets onto a sheet of paper. Think about other ways to model distances such as on the chalkboard, sidewalk, school yard, or cash register tape.
- If a model you make uses distance that would put some planets out of the schoolyard, you can use a map to show that part of your model.
- Try several different scales. Do the mathematical calculations for the largest and smallest planets, and think about how you would model the sizes that you have calculated. Your scale must be able to reasonably demonstrate both the small and large sizes. Your distances and planet diameters do not have to be expressed in kilometers. You may want to try converting the units to meters or centimeters.
- To show a distance that may extend beyond the boundary of your model, you can label an arrow or a dotted line at the outer limit with the additional distance the planet extends from that boundary.
- If you decide to use a scale that creates a model that is too large to bring into the classroom, you might consider videotaping it. You could show the video as part of your class presentation.

## ◆ Project Time Line

| Task | Due Date | Teacher's Initials |
|---|---|---|
| **1.** Scale for distances between planets selected and approved | __________ | __________ |
| **2.** Model of distances between planets completed | __________ | |
| **3.** Scale for planet sizes selected and approved | __________ | __________ |
| **4.** Model of planet sizes completed | __________ | |
| **5.** Scale for combination model selected and approved | __________ | |
| **6.** Combination model approved | __________ | __________ |
| **7.** Class presentation completed | __________ | |

Name ______________________ Date __________ Class __________

CHAPTER 2 PROJECT WORKSHEET 1

2

# Separate Distance and Diameter Models

## ◆ Distance Between Planets

*Complete the following data table. When it has been completed, you are ready to get your teacher's approval and begin modeling the distances between the planets.*

| Solar System Object | Radius of Orbit (km) | 1:10,000,000,000 Scale | 1:50,000,000,000 Scale | Your Scale (1:________) |
|---|---|---|---|---|
| Sun | Center | Center | Center | Center |
| Mercury | 58,000,000 | 5.8 m | 1.2 m | |
| Venus | 108,000,000 | 10.8 m | 2.2 m | |
| Earth | 150,000,000 | | | |
| Mars | 228,000,000 | | | |
| Jupiter | 778,000,000 | | | |
| Saturn | 1,427,000,000 | | | |
| Uranus | 2,871,000,000 | | | |
| Neptune | 4,497,000,000 | | | |
| Pluto | 5,913,000,000 | | | |

## ◆ Planet Size

*Complete the following data table. When it has been completed, you are ready to get your teacher's approval and begin modeling the sizes of planets.*

| Solar System Object | Diameter (km) | 1:1,000,000,000 Scale | 1:500,000,000 Scale | Your Scale (1:________) | Model Object |
|---|---|---|---|---|---|
| Sun | 1,392,000 | 1.39 m | 2.78 m | | |
| Mercury | 4,878 | 5 mm | 10 mm | | |
| Venus | 12,104 | | | | |
| Earth | 12,756 | | | | |
| Mars | 6,794 | | | | |
| Jupiter | 142,800 | | | | |
| Saturn | 120,540 | | | | |
| Uranus | 51,200 | | | | |
| Neptune | 49,500 | | | | |
| Pluto | 2,200 | | | | |

# Combination Model

*Complete the following data table. Make a second data table like the one below in which you can use your own choice of scale. When it has been completed, you are ready to get your teacher's approval and begin the combination model.*

| Solar System Object | Diameter (km) | 1:10,000,000,000 Scale | Distance from Sun (km) | 1:10,000,000,000 Scale | Model Object |
|---|---|---|---|---|---|
| Sun | 1,392,000 | 13.9 cm | Center | Center | |
| Mercury | 4,878 | 0.5 mm | 58,000,000 | 5.8 m | |
| Venus | 12,104 | | 108,000,000 | | |
| Earth | 12,756 | | 150,000,000 | | |
| Mars | 6,794 | | 228,000,000 | | |
| Jupiter | 142,800 | | 778,000,000 | | |
| Saturn | 120,540 | | 1,427,000,000 | | |
| Uranus | 51,200 | | 2,871,000,000 | | |
| Neptune | 49,500 | | 4,497,000,000 | | |
| Pluto | 2,200 | | 5,913,000,000 | | |

## ◆ Analyzing and Presenting

*Complete the following tasks on a separate sheet of paper. When they have been completed, you are ready to plan your presentation.*

**1.** Write a paragraph summarizing your experience with model building. What scales did you choose? Why did you choose these scales? What problems did you encounter in model building?

**2.** Write a paragraph discussing what you have learned about the size of the solar system. What did you notice about the spacing of the planets? Are some closer together than others?

**3.** Write a paragraph about scaling. What are the advantages to building models to scale? What are the disadvantages of building models to scale?

CHAPTER 2

# Model of the Solar System

In evaluating how well you complete the Chapter 2 Project, your teacher will judge your work in four categories. In each, a score of 4 is the best rating.

| | 4 | 3 | 2 | 1 |
|---|---|---|---|---|
| ***Mathematical Calculations*** | Accurately computes diameters and distances for all models. | Accurately computes most diameters and distances for the models. | Accurately computes some diameters and distances for the models. | Accurately computes only a few diameters and distances for the models. |
| ***Model Scale Selection*** | All the scales chosen are useful for demonstrating planet diameters and distances. All objects chosen are the correct size for the scale. | Two of the scales chosen are useful for demonstrating planet diameters and distances. Most objects chosen are the correct size for the scale. | One of the scales chosen are useful for demonstrating planet diameters and distances. Some objects chosen are the correct size for the scale. | Scales for the models are unreasonable and not useful for illustrating planet diameters and distances. Objects chosen are not the correct size for the scale. |
| ***Understanding Size and Scaling*** | The student completely describes size and scale on the worksheets and in class discussions. | The student adequately describes size and scale on the worksheets and in class discussions. | The student somewhat describes size and scale on the worksheets and in class discussions. | The student minimally describes size and scale on the worksheets and in class discussions. |
| ***Class Presentation*** | The student makes a thorough, well-organized presentation. Student explains why particular scales for models were chosen. | The student makes an adequate presentation. Student partially explains why particular scales for the models were chosen. | Presentation is somewhat hard to follow. Very little explanation given for choice of particular scales. | Presentation is very brief and hard to follow. No explanation given for choice of particular scales. |

# Observing the Solar System (pp. 50–55)

## Objectives

After completing the lesson, students will be able to

- explain how the heliocentric and geocentric models of the solar system differ;
- explain what Kepler discovered about the orbits of the planets;
- identify two factors that keep the planets in their orbits.

## 1 Engage/Explore

- **Activating Prior Knowledge** While showing students a photograph of the night sky, reinforce the idea that the objects in our solar system are moving.
- **Discover** Allow students to use a toy truck and rocks to examine how mass and speed affect an object's motion.

## 2 Facilitate

- **Building Inquiry Skills: Inferring** Guide students in evaluating why the ancient astronomers only could see five planets.
- **Cultural Diversity** Tell students that other cultures gave objects in the sky names meaningful in those cultures. Ask students to give examples of such names from other cultures.
- **Using the Visuals: Figure 1** Ask students to infer why exposing the film for several hours produced the circular effect shown in the figure.
- **Building Inquiry Skills: Comparing and Contrasting** Encourage students to diagram geocentric and heliocentric views of the solar system.
- **Including All Students** Invite students to research and draw diagrams of the phases of Venus.
- **Building Inquiry Skills: Modeling** Have students make a living orrery.
- **Building Inquiry Skills: Inferring** After diagramming a planet's elliptical orbit, challenge students to infer why scientists record a planet's average distance from the sun.
- **Try This** Have students draw an ellipse using pushpins and string. Challenge them to examine how changing the distance between the pushpins affects the ellipse's shape.
- **Using the Visuals: Figure 5** Lead students to see that the planet would continue to move in a straight path if the sun's gravity disappeared.
- **Addressing Naive Conceptions** Using a bicycle wheel and nine pieces of tape, demonstrate how eclipses occur.
- **Real Life Learning** Discuss planetary alignments and sensationalist books and newspapers.
- **Building Inquiry Skills: Graphing** Direct students to make a bar graph using the data in the figure.

## 3 Assess

- **Section 1 Review** Check students' understanding of section content by having them answer the section review questions.
- **Performance Assessment** Assess students' understanding of the motion of the planets by having them create concept maps that illustrate the relationships among Kepler, Ptolemy, Newton, Galileo, heliocentric, elliptical orbit, geocentric, inertia, and gravity.

2

Name ______________________ Date __________ Class ______________

SECTION 2-1 SECTION SUMMARY

# Observing the Solar System

**Guide for Reading**

- **How do the heliocentric and geocentric descriptions of the solar system differ?**
- **What did Kepler discover about the orbits of the planets?**
- **What two factors keep the planets in their orbits?**

2

Ancient Greek astronomers thought that Earth was inside a rotating dome called the celestial sphere. Since *geo* is the Greek word for Earth, an Earth-centered explanation is known as a **geocentric** system. **In a geocentric system, Earth is at the center of the revolving planets.** In A.D. 140, the Greek astronomer Ptolemy made a new model of the motion of the planets. Like earlier Greek astronomers, Ptolemy thought Earth was at the center of the system of planets. In Ptolemy's model, however, the planets move on little circles that move on bigger circles. He was trying to explain why the direction and speed of planets seem to change as they move across the sky.

In the early 1500s, the Polish astronomer Nicolaus Copernicus developed a new model for the motions of planets. His sun-centered system is called **heliocentric.** *Helios* is Greek for "sun." **In a heliocentric system, Earth and the other planets revolve around the sun.**

The Italian astronomer Galileo Galilei lived nearly 100 years after Copernicus. He made two discoveries supporting the heliocentric model. First, he saw four moons revolving around Jupiter. Second, he saw that Venus goes through phases similar to those of Earth's moon. A heliocentric model explains both discoveries.

Copernicus thought that the planets' orbits were circles. He based his conclusions on observations made by the ancient Greeks. In the late 1500s, Tycho Brahe made more accurate observations of the planet's orbits. Johannes Kepler analyzed Brahe's data. **Kepler had discovered that the orbit of each planet is an ellipse.** An **ellipse** is an elongated circle, or oval shape. However, he could not explain why the planets stayed in orbit.

The English scientist Isaac Newton provided the answer. **Newton concluded that two factors—inertia and gravity—combine to keep the planets in orbit.** A moving object will continue to move until a force acts to stop its motion. This tendency of a moving object to continue in a straight line or of a stationary object to remain in place is the object's **inertia.** Gravity is the force that attracts all objects toward each other. Newton figured out that Earth keeps pulling the moon toward it with gravity, while the moon keeps moving ahead because of its inertia. Earth curves away as the moon falls toward it, so the moon ends up in orbit. In the same way, the planets end up in orbit around the sun.

Name ______________________ Date __________ Class ______________



2

# The Sun

## ◆ Understanding Main Ideas

*Label the diagram of the sun below.*

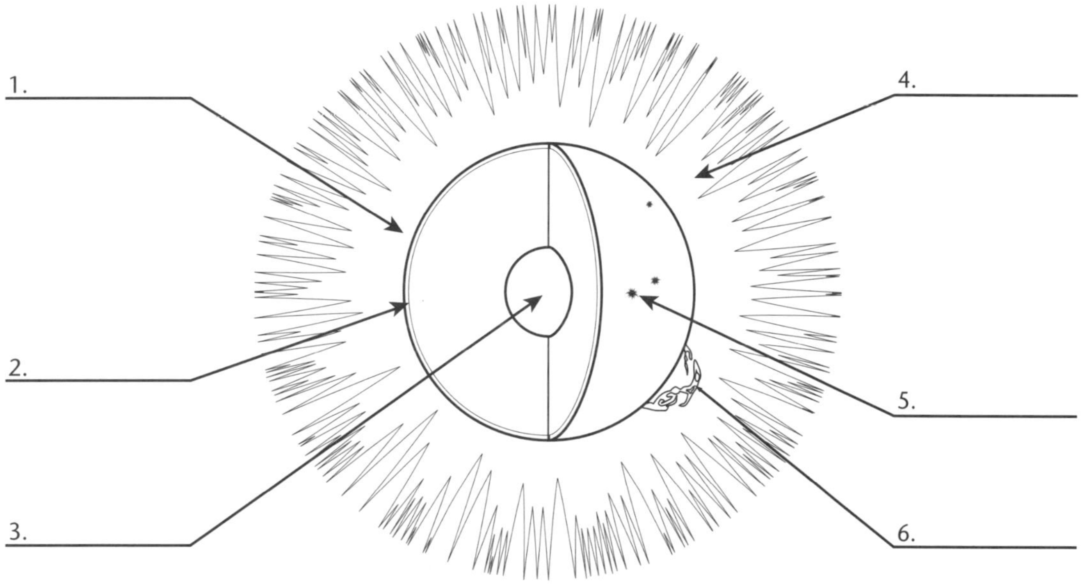


## ◆ Building Vocabulary

Match each term with its definition by writing the letter of the correct definition in the right column on the line beside the term in the left column.

_______ **7.** solar flare

_______ **8.** core

_______ **9.** chromosphere

_______ **10.** sunspot

_______ **11.** corona

_______ **12.** nuclear fusion

_______ **13.** photosphere

_______ **14.** solar wind

**a.** the layer of the sun's atmosphere that makes light

**b.** the layer of the sun's atmosphere that has a reddish glow

**c.** the layer of the sun's atmosphere that looks like a halo during an eclipse

**d.** areas of gas on the sun that are cooler than the gases around them

**e.** reddish loops of gas that link sunspots in pairs or groups

**f.** explosions that occur when the loops between sunspots suddenly connect

**g.** a stream of particles produced by the corona

**h.** the center of the sun

**i.** the joining of hydrogen atoms to form helium

Name ______________________ Date __________ Class ______________

SECTION 2-2 ENRICH

# Sunspot Clues

Although sunspots were once unexplained blemishes on the sun's surface, their behavior has helped to solve some of the sun's mysteries. One of these mysteries dealt with the rotation of the sun. Did it rotate, as did other large objects in the solar system? And, if it did rotate, what was its period of rotation? Astronomers helped to answer these questions by observing the behavior of sunspots.

2

The diagrams below show how a series of sunspots behave over time.

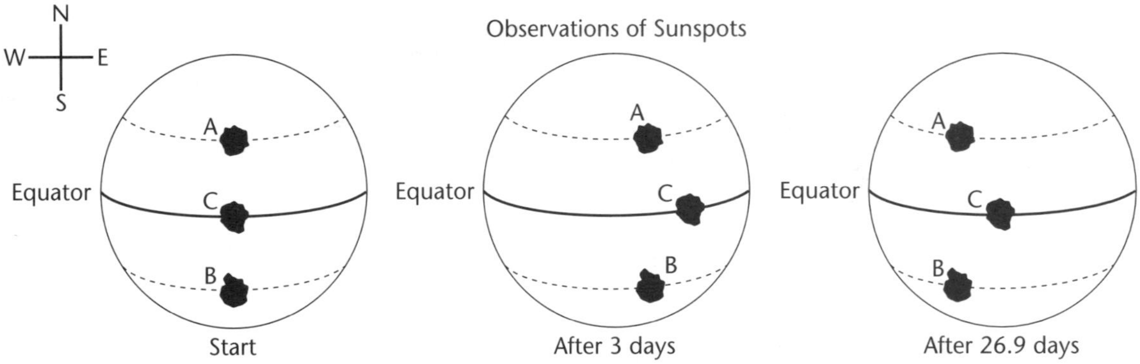


*Answer the questions below on the back of this sheet or on a separate sheet of paper.*

**1.** Why have astronomers inferred that the sun rotates?

**2.** In what direction does the sun rotate?

**3.** Sunspots at the equator take 26.9 days to move once around the sun. What can you infer about how long the sunspots A and B take to move around the sun, compared to sunspot C, on the equator?

**4.** Why must astronomers say that the sun rotates once every 27 to 31 days rather than give an exact number?

# The Inner Planets (pp. 62–69)

## Objectives

After completing the lesson, students will be able to

- ◆ describe the main characteristics of the inner planets.

## 1 Engage/Explore

- ◆ **Activating Prior Knowledge** Show students a colored map of the solar system and have them point out Earth and other planets they think are most like Earth.
- ◆ **Discover** Have students model what it's like to look at Mars from Earth.

## 2 Facilitate

- ◆ **Building Inquiry Skills: Interpreting Data** Help students interpret the data in Figure 10 by asking them questions comparing the planets.
- ◆ **Integrating Earth Science** Invite students to use hard-boiled eggs to model layers of Earth.
- ◆ **Building Inquiry Skills: Observing** Encourage students to model the difficulty making observations of Mercury using a coin held in front of a light bulb.
- ◆ **Using the Visuals: Figure 12** Have students describe the features shared by the moon and Mercury.
- ◆ **Demonstration** Invite students to identify differences between the rotations of Venus and Earth while demonstrating their rotations.
- ◆ **Sharpen Your Skills** Using the data in Figure 10, challenge students to make a line graph of the average distance from the sun and period of revolution of Mercury, Venus, Earth, and Mars.
- ◆ **Including All Students** To help students understand the origin of the term greenhouse effect, show them a picture of a greenhouse with plants growing inside and explain its use.
- ◆ **Music Connection** Encourage students to listen for astronomical connections while listening to recordings of this work.
- ◆ **Using the Visuals: Figure 14** Have students infer why scientists used radar to get an image of the volcano on Venus. Ask students why an astronomer might use an exaggerated scale when examining an image.
- ◆ **Addressing Naive Conceptions** Direct students to prepare a Fact/Fiction sheet to distinguish scientific findings about Mars from science fiction.
- ◆ **Try This** Allow students to discover how hard it is to explore another planet by remote control. Provide paper covered goggles for the students to use.
- ◆ **Building Inquiry Skills: Observing** To help students determine the effects of flowing water on land, have them slowly pour water over mounds of moist sand.

## 3 Assess

- ◆ **Section 3 Review** Check students' understanding of the inner planets by having them answer the section review questions.
- ◆ **Performance Assessment** Assess students' understanding of the main characteristics of the inner planets by having them create models of the four inner planets using art supplies and classroom items.

2

Name ______________________ Date __________ Class __________

SECTION 2-3 SECTION SUMMARY

# The Inner Planets

**Guide for Reading**

- What are the main characteristics of the inner planets?

2

Mercury, Venus, Earth, and Mars are more similar to each other than they are to the five outer planets. **The four inner planets are small and have rocky surfaces.** These planets are often called the **terrestrial planets,** from the Latin word *terra,* or "earth."

Earth's atmosphere includes the gases nitrogen, oxygen, and a small amount of carbon dioxide, as well as water vapor and clouds of water droplets. About 70 percent of Earth is covered with water. Earth is made up of three layers: a crust, a mantle, and a core. The crust includes the solid rocky surface. Under the crust is the mantle, a layer of molten rock. Under the mantle is the core. The outer core is liquid, but the inner core is probably solid and made up of iron and nickel.

Mercury is the planet closest to the sun. Because of this, people from Earth never get a good view of Mercury. Mercury is smaller than Earth's moon and has no moons of its own. The planet's interior is probably made of iron and nickel, and its surface has many plains and craters. Because the planet is so close to the sun, the side facing the sun reaches temperatures of 450°C. However, the temperature drops to –170°C at night.

Venus is similar in size to Earth. Astronomers think the density and internal structure of Venus are also similar to Earth's. Venus rotates so slowly that its "day" is longer than its "year." Venus rotates from east to west, the opposite direction from most other planets and moons. This type of "backward" rotation is called **retrograde rotation.** The atmosphere has 90 times the air pressure of Earth's atmosphere. The atmosphere is mostly carbon dioxide with clouds partly made up of sulfuric acid. The carbon dioxide in the planet's atmosphere traps the sun's heat, causing the surface temperature of Venus to be about 460°C. This trapping of heat by the atmosphere is called the **greenhouse effect.** Explorations by the *Magellan* probe in 1991 showed that Venus has active volcanoes, many craters, and strange domes not found on other planets.

Mars is called the "red planet." Its surface is covered in red dust that is blown around in windstorms. Astronomers have found water on Mars in the form of ice at the north pole. A layer of frozen carbon dioxide covers this ice in the winter. In 1997, photographs from orbit showed evidence that water flowed millions of years ago. The planet Mars has an atmosphere that is mostly carbon dioxide and has only 1 percent the pressure of Earth's atmosphere. Like Earth, Mars is tilted on its axis so the seasons change. Mars has two very small moons, Phobos and Deimos.

Name ______________________ Date __________ Class __________

SECTION 2-3 **REVIEW AND REINFORCE**

# The Inner Planets

## ◆ Understanding Main Ideas

*Label the diagram with the names of the inner planets.*

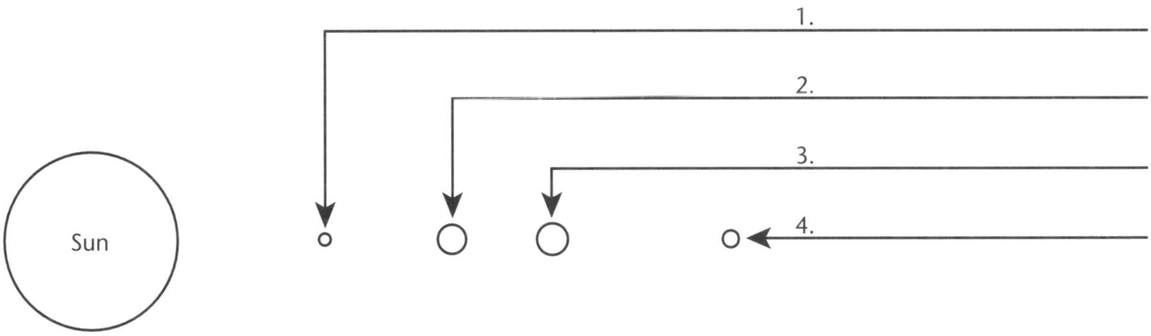


*Write the planet or planets the statement describes.*

______________ **5.** has a rocky surface

______________ **6.** 70 percent is covered with water

______________ **7.** rotates in the opposite direction from most other planets and moons

______________ **8.** called the "red planet" because of the color of the dust

______________ **9.** has at least one moon

______________ **10.** similar to each other in size, density, and internal structure

______________ **11.** has almost no atmosphere

______________ **12.** atmosphere so heavy and thick that it would crush a human

______________ **13.** has a tilted axis so that the planet has seasons

______________ **14.** atmosphere has low air pressure and is mostly carbon dioxide

## ◆ Building Vocabulary

*Write a definition for each of the following terms.*

**15.** terrestrial planets

______________________________

**16.** retrograde rotation

______________________________

**17.** greenhouse effect

______________________________

2

2

# Atmospheres of Earth and Venus

Earth and Venus are about the same size. Still, it is hard to imagine how the conditions on the surfaces of Earth and Venus could be more different. Earth's surface is cool, while that of Venus is 460°C. You have learned that this is because of the greenhouse effect caused by the carbon dioxide gas in Venus's atmosphere. Earth has a smaller greenhouse effect because there is much less carbon dioxide in its atmosphere. Why is this so, if the two planets are so similar in other ways?

Both Earth and Venus get their carbon dioxide from the same source: erupting volcanoes. Unlike on Venus, Earth's carbon dioxide is constantly removed from the atmosphere by two things that Venus lacks: liquid water and green plants. Carbon dioxide dissolves easily in water. Once it is there, it soon joins with other chemicals to form solid substances that become part of sea-floor rocks. This carbon dioxide is trapped unless the rock melts. As a result, much of Earth's carbon dioxide is "locked up" in these rocks. Plants also absorb carbon dioxide. Thus, some of the gas is locked up in their bodies. On Venus, carbon dioxide simply builds up in the atmosphere.

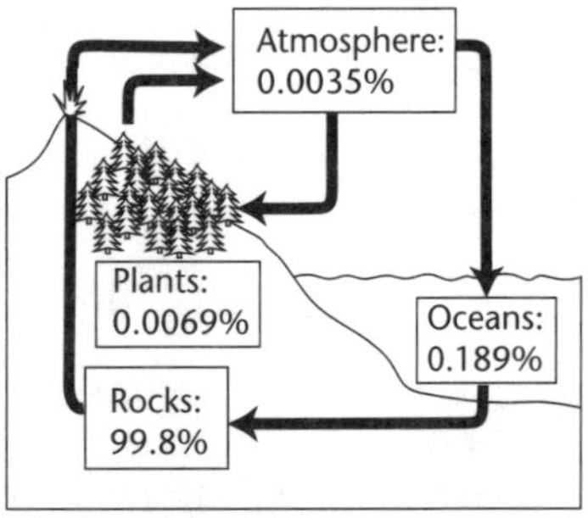


Location and pathway of carbon dioxide at or near Earth's surface

Figure 1

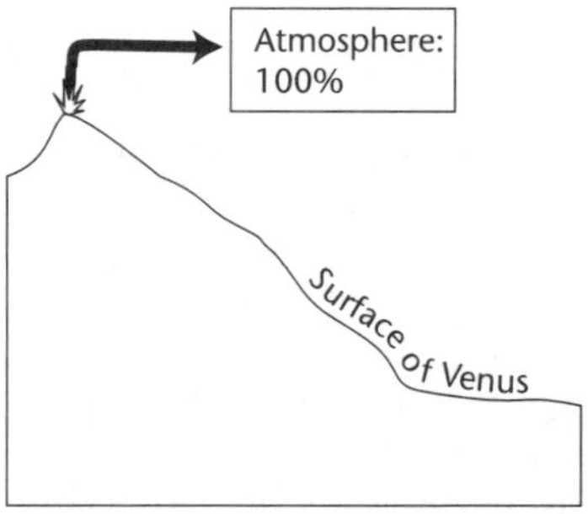


Location and pathway of carbon dioxide at or near the surface of Venus

Figure 2

*Use the diagrams to answer the questions below on the back of this sheet or on a separate sheet of paper.*

1. Figure 1 shows, in very simple terms, the pathways taken by carbon dioxide on Earth. According to the diagram, where does carbon dioxide go when it leaves the atmosphere?
2. On Earth, where is most of the carbon dioxide located at any one time? On Venus, where is most of the carbon dioxide located at any one time?
3. On Earth, how does carbon dioxide move from the atmosphere into rocks? Why does this movement not occur on Venus?
4. If the temperature of Earth's surface were to increase by 100°C, what would happen to the amount of carbon dioxide in the atmosphere? Why?
5. If all carbon dioxide coming out of volcanoes were to remain in the atmosphere, what further effect would this have on Earth's temperature?

# The Outer Planets (pp. 70–79)

## Objectives

After completing the lesson, students will be able to

- ◆ describe the main characteristics of the gas giant planets;
- ◆ explain how Pluto differs from the other outer planets.

## 1 Engage/Explore

- ◆ **Activating Prior Knowledge** Assign an outer planet to each student. Have them write three things they think they know about their planet.
- ◆ **Discover** Using a quarter as Earth's diameter, invite students to draw a circle of the other planets relative to the size of Earth.

## 2 Facilitate

- ◆ **Using the Visuals: Figure 19** Have students evaluate the orbits of each planet as they trace the orbits. Ask them questions about the distances the planets are from the sun.
- ◆ **Building Inquiry Skills: Applying Concepts** Invite students to imagine sending a space probe to a gas giant. Ask them to determine what kind of landing gear the space probe will need.
- ◆ **Building Inquiry Skills: Making Models** Allow students to model the motion of the Great Red Spot and to infer what forces are causing it to swirl.
- ◆ **Demonstration** Demonstrate how a large planet such as Saturn can float using a large polystyrene sphere and a small marble.
- ◆ **Try This** Provide students the opportunity to build a scale model of Saturn.
- ◆ **Using the Visuals: Figure 23** Have students describe the appearance of each of the moons shown in the figure.
- ◆ **Building Inquiry Skills: Comparing and Contrasting** Students compare and contrast Jupiter and Saturn.
- ◆ **Visual Arts Connection** Show students pointillist paintings and compare them to Saturn's rings.
- ◆ **Using the Visuals: Figure 24** Have students lay a pencil over the axis of rotation then trace the orbit of Uranus by moving the pencil.
- ◆ **Language Arts Connection** Describe the gods each planet was named for and ask students to explain why each god was chosen.
- ◆ **Building Inquiry Skills: Relating Cause and Effect** Invite students to predict how a large object's gravity could affect the orbit of a planet. Allow them to model this using a marble placed on a piece of spandex stretched over a bowl.
- ◆ **Building Inquiry Skills: Classifying** Challenge students to explain why Pluto should be classified as a planet, an asteroid, or as some other object.

## 3 Assess

- ◆ **Section 4 Review** Check students' understanding of section content by having them answer the section review questions.
- ◆ **Performance Assessment** Assess students' understanding of the outer planets by having them draw diagrams of the outer planets that show their relative sizes, orders, and a distinguishing characteristic.

## Lab

- ◆ **Skills Lab: Developing Hypotheses** Challenge students to discover how a planet's distance from the sun affect its period of revolution.

Name ______________________ Date __________ Class __________

SECTION 2-4 SECTION SUMMARY

# The Outer Planets

**Guide for Reading**

- What are the main characteristics of the gas giant planets?
- How is Pluto different from the other outer planets?

2

**The first four outer planets—Jupiter, Saturn, Uranus, and Neptune—are much larger than Earth, and are made almost entirely of gases.** Because these four planets are all so large, they are also called the **gas giants.** The fifth planet, Pluto, is small and rocky like the terrestrial planets.

Because the gas giants have so much mass, they have a stronger gravitational force than the terrestrial planets. This prevents their gases from escaping. The composition of their atmospheres is about 75 percent hydrogen, 24 percent helium, and 1 percent other elements. None of the giant planets has a solid surface, and astronomers think that each has a partly solid core.

Jupiter is the most massive planet. Its atmosphere appears as colorful bands of thick clouds with a Great Red Spot that is many times bigger than Earth. The Great Red Spot is an area of swirling clouds that appears to be an ongoing storm. Jupiter has 17 moons.

Saturn is the second largest planet in the solar system. If you include its rings, it has a larger overall diameter than Jupiter. Saturn is the least dense planet in the solar system. The hundreds of rings around Saturn are made of chunks of ice and rock, each in orbit around Saturn. Rings have also been discovered around the other gas giants. Saturn has 19 moons.

The English astronomer William Herschel discovered Uranus in 1781. It is about four times the diameter of Earth and is twice as far from the sun as Saturn. Uranus looks bluish because of traces of methane gas in its atmosphere. Uranus's axis is 90° off vertical. It rotates from top to bottom. Astronomers have discovered 17 moons around Uranus.

Neptune was discovered in 1846 when astronomers noticed that Uranus was not quite following the orbit they had predicted. They hypothesized that the gravity of an unseen planet was affecting Uranus's orbit. This planet was Neptune. Neptune is 30 times Earth's distance from the sun. Its atmosphere contains visible clouds. Astronomers have discovered 8 moons revolving around Neptune.

Clyde Tombaugh discovered Pluto in 1930 while searching for a large object that might be affecting Neptune's orbit. Pluto has a single moon, Charon. Pluto and Charon have solid surfaces and masses much less than that of Earth. Since Charon is more than half the size of Pluto, astronomers often consider them to be a double planet. Pluto and Charon revolve around the sun only once every 249 Earth years.

Name ____________________ Date __________ Class ____________

**SECTION 2-4** **REVIEW AND REINFORCE**

# The Outer Planets

## ◆ Understanding Main Ideas

*Answer the following questions. Use the back of this sheet if you need more space.*

**1.** What are the five outer planets?

______________________________

**2.** Which planets are the gas giants?

______________________________

**3.** What are the two main differences between Pluto and the gas giants?

______________________________

**4.** Why doesn't the gas on a gas giant escape into space, as it has on Mercury?

______________________________

**5.** What object in the solar system has a composition similar to that of the gas giants?

______________________________

**6.** What planet is by far the most massive of all the planets that revolve around the sun?

______________________________

**7.** What are Saturn's rings made of?

______________________________

**8.** Why did astronomers know where to look to discover Neptune?

______________________________

**9.** Why do astronomers sometimes consider Pluto and its moon, Charon, to be a double planet?

______________________________

## ◆ Building Vocabulary

**10.** Define *gas giant.*

______________________________

2

Name ______________________ Date __________ Class ______________

SECTION 2-4 ENRICH

# The Orbits of Neptune and Pluto

The orbit of Neptune is very close to circular. Pluto's orbit is more elliptical.

The diagram below shows the orbits of Pluto and Neptune. The unit of distance used in the diagram is the *astronomical unit,* or a.u. An a.u. is the average distance from Earth to the sun, about 150 million kilometers. Neptune's distance from the sun is 30 astronomical units or 4,497 million kilometers.

2

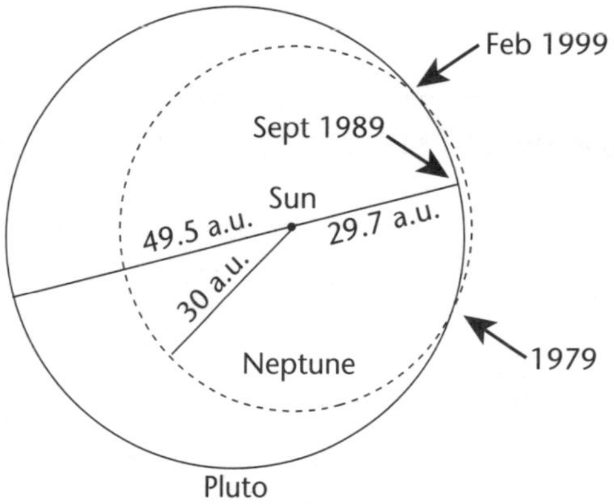


*Answer the following questions on a separate sheet of paper.*

**1.** In 1989, Pluto was as close to the sun as it ever gets. How close was it?

**2.** What was Neptune's distance from the sun in 1989?

**3.** Pluto takes 249 years to complete one revolution around the sun. When next will Pluto and Neptune be an equal distance from the sun?

**4.** When will Pluto next be as far from the sun as if ever gets?

**5.** Is one planet farther from the sun now? If so, which one?

# Comets, Asteroids, and Meteors (pp. 80–83)

## Objectives

After completing the lesson, students will be able to

- describe the characteristics of comets and asteroids;
- identify where meteoroids come from.

## 1 Engage/Explore

- **Activating Prior Knowledge** Invite students to discuss what a shooting star is. Tell students that in this section, they investigate the nature of comets, asteroids, and shooting stars, which are actually meteors.
- **Discover** Allow students to discover what determines which way the tail of a comet points. They can model this with a ball of clay, string, and an electric fan.

## 2 Facilitate

- **Using the Visuals: Figure 28** As students study the figure, ask them how the orbit of a comet is different from the orbit of a planet.
- **Building Inquiry Skills: Inferring** As students explore the structure of a comet, have them make inferences about observing comets from Earth.
- **Building Inquiry Skills: Predicting** Ask students to predict whether they are likely to ever have the opportunity to see Halley's comet. Guide them in calculating when Halley's comet will next be visible from Earth.
- **Integrating Earth Science** Challenge students to locate the Yucatan peninsula on a world map. Guide them in understanding that asteroids and other objects from space that land on Earth become a part of the rock cycle.
- **Building Inquiry Skills: Hypothesizing** Allow students to use liquid plaster of Paris to see how meteorites form craters and how objects of different sizes form craters of different sizes.

## 3 Assess

- **Section 5 Review** Check students' understanding of comets, asteroids, and meteors by having them answer the section review questions.
- **Performance Assessment** Assess students' understanding of the characteristics of asteroids by having them narrate the life story of an asteroid that orbits close to Earth and eventually hits the ground.

2

# Comets, Asteroids, and Meteors

**Guide for Reading**

- **What are the characteristics of comets and asteroids?**
- **Where do meteoroids come from?**

2

The sun, planets, and moons aren't the only objects in the solar system. There are also millions of smaller objects, most of which are classified as comets and asteroids.

You can think of a **comet** as a "dirty snowball" about the size of an Earth mountain. **Comets are chunks of ice and dust whose orbits are usually very long, narrow ellipses.** Because their orbits are so elliptical, few comets pass near Earth. When a comet gets close enough to the sun, the ice turns into gas, releasing dust. The gas and dust form an outer layer called the *coma.* The inner layer of the comet is called the nucleus. The coma and the nucleus make up the comet's head, which is the brightest part of the comet. Some of the gas from a comet is pushed away from the sun by the solar wind. This forms the comet's tail. A comet's tail is stretched very thinly and can be hundreds of millions of kilometers long.

In 1705, the English astronomer Edmond Halley calculated that several observed comets had the same orbit. Halley suggested they were actually the same comet. He calculated that this comet appeared about every 76 years and predicted that it would reappear in 1758. When his prediction came true, the comet was named Halley's Comet.

Between 1801 and 1807, astronomers discovered four small objects between Mars and Jupiter. Over the next 80 years, they found 300 more. These objects, called **asteroids,** are too small and too numerous to be considered full-fledged planets. **Most asteroids revolve around the sun between the orbits of Mars and Jupiter.** This region of the solar system is known as the **asteroid belt.**

Astronomers have discovered more than 10,000 asteroids. Some come near Earth's orbit. An asteroid hit Earth 65 million years ago. It exploded, making a crater 200 kilometers in diameter in the Yucatan Peninsula of Mexico. The explosion probably started huge fires that destroyed much of Earth's forests and grass. Scientists hypothesize that this caused the extinction of many types of animals, including dinosaurs.

A **meteoroid** is a chunk of rock or dust in space. **Meteoroids usually come from comets or asteroids.** When a meteoroid enters the Earth's atmosphere, friction makes it burn up and produce a streak of light you see in the sky—a **meteor.** If a meteoroid is large enough, it may not burn up completely. Meteoroids that pass through the atmosphere and hit Earth's surface are called **meteorites.** Meteorites fall all over Earth. Most of them look like stones. A few are made almost entirely of iron and nickel and are unusually heavy for their size. This makes them more likely to be identified as meteorites than as Earth rocks.

Name ______________________ Date __________ Class __________

SECTION 2-5 REVIEW AND REINFORCE

# Comets, Asteroids, and Meteors

## ◆ Understanding Main Ideas

*Complete the following table.*

| Object | Description | Location/Movement |
|---|---|---|
| Asteroid | | |
| Comct | | |
| Meteoroid | | |

*Answer questions 1 through 3 on a separate sheet of paper.*

**1.** Explain what happens to a meteoroid in order for it to become a meteorite.

**2.** Describe these parts of a comet: head, nucleus, coma, tail.

**3.** How can you tell a meteor from a comet?

## ◆ Building Vocabulary

*From the list below, choose the term that best completes each sentence.*

| | | |
|---|---|---|
| asteroid | asteroid belt | comet |
| meteor | meteoroid | meteorite |

**4.** When a meteoroid enters Earth's atmosphere, friction causes it to burn up and produce a streak of light called a(n) ________________.

**5.** A chunk of ice and dust whose orbit is usually a long narrow ellipse is a(n) ________________.

**6.** If a meteoroid hits Earth's surface, it is called a(n) ________________.

**7.** An object that revolves around the sun, but is too small to be considered a planet, is a(n) ________________.

**8.** A chunk of rock or dust in space that usually comes from a comet or an asteroid is called a(n) ________________.

**9.** The region of the solar system between the orbits of Mars and Jupiter is known as the ________________.

Name ______________________ Date __________ Class ______________

SECTION 2-5 ENRICH

# Evidence of a Large Meteorite

Meteoroids frequently hit Earth's atmosphere. If you were to get away from city lights on a clear night and patiently look at the sky for an hour or two, you would probably see several meteors, visual evidence that meteoroids are burning up due to air friction as they pass through Earth's atmosphere.

2

If Earth didn't have an atmosphere, its surface would probably look like the surface of the moon. Earth's atmosphere not only prevents many meteoroids from hitting the surface, it also tends to erase and cover up craters over time through the action of wind and rain. Scientists must look for other evidence that large meteorites have hit Earth. One place they look is in rocks.

When a large meteorite hits Earth, the result is similar to a huge bomb exploding. Some of the material from the meteor turns to dust that goes high into the atmosphere where winds carry it over the entire surface of Earth. Over millions of years, mud, sand, and dust (including dust from meteors) can build up and eventually turn into layers of rock.

Iridium is an element that is very rare in rocks on Earth, except very deep below the surface. It is much more common in rocks from space. The graph below shows the amount of iridium present in some rocks at a site in Italy.

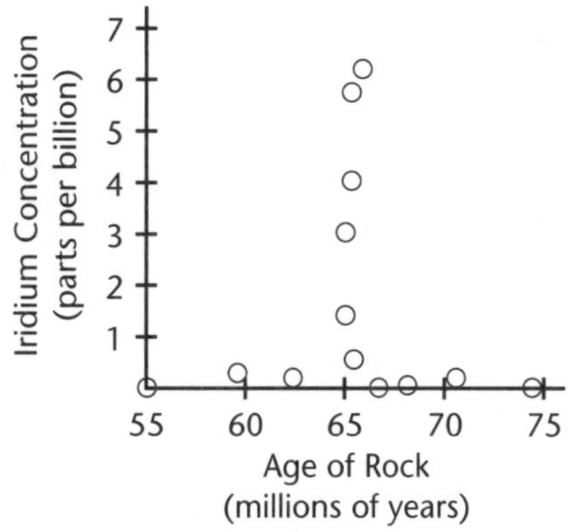


*Study the graph and answer the following questions on the back of this sheet or on another sheet of paper.*

1. How are the moon's craters evidence that large meteoroids probably hit Earth?
2. Why is evidence of meteoroid impact rare on Earth, despite the fact they often pass into Earth's atmosphere?
3. About what age are the rocks that show the highest level of iridium?
4. What are two possible sources for the high iridium level shown in the graph?
5. In several other parts of the world, rocks of the same age have similar levels of iridium. What can you infer from this information?

*Integrating Life Science*

# Is There Life Beyond Earth? (pp. 84–88)

## Objectives

After completing the lesson, students will be able to

- describe the conditions living things need to exist on Earth;
- explain why scientists think Mars and Europa are good places to look for signs of life.

## 1 Engage/Explore

- **Activating Prior Knowledge** Explain that living things as we know them all need certain things in order to survive.
- **Discover** Allow students to observe yeast in order to determine if yeast is alive or not.

## 2 Facilitate

- **Cultural Diversity** Encourage students to research NASA's web site to find out what messages were sent aboard *Voyager.* Have them list the things that they would send to reflect their own culture.
- **Sharpen Your Skills** Have students explain in a letter to a friend from another planet why the conditions on Earth make it the ideal place for living things.
- **Addressing Naive Conceptions** Ensure that students understand that the search for extraterrestrial life is not only a search for intelligent life forms.
- **Building Inquiry Skills: Drawing Conclusions** Ask students what hypothesis scientists were testing when they sent the *Viking* landers to Mars and what samples they were examining.
- **Building Inquiry Skills: Making Judgements** Allow teams of students to debate the existence of life on Mars.
- **Including All Students** Encourage students to speculate on what kinds of life forms could exist beneath Europa's icy crust.
- **Language Arts Connection** Students are encouraged to read stories about life on other planets and separate fact from fiction.

## 3 Assess

- **Section 6 Review** Check students' understanding of life beyond Earth by having them answer the section review questions.
- **Performance Assessment** Assess students' understanding of life on other planets by having them explain whether each of the nine planets in our solar system would be good places to search for signs of life.

## Science and Society

- **Space Exploration—Is It Worth the Cost?** Direct students to list the costs and benefits of space exploration. Then have students make a chart of three different approaches to space exploration and list the benefits and drawbacks of each approach.

Name ________________________________ Date __________ Class ________________

SECTION 2-6 SECTION SUMMARY

# Is There Life Beyond Earth?

**Guide for Reading**

- What conditions do living things need to exist on Earth?
- Why do scientists think Mars and Europa are good places to look for signs of life?

2

All living things on Earth have several characteristics in common. Living things are made up of cells. Living things take in energy and use it to grow and develop. They reproduce and give off waste.

**Earth has liquid water and a suitable temperature range and atmosphere for living things to survive.** These are sometimes called the "Goldilocks conditions." Are these conditions necessary for life? Or are they just conditions that Earth's living things happen to need? Scientists only have life forms on Earth to study. Until they find life somewhere else, there is no way to answer these questions. Life other than that on Earth would be called **extraterrestrial life.**

On Earth, scientists have recently discovered life forms that live outside the range of conditions they once thought necessary to support life. There are giant tube worms that live in the deep ocean at a very high pressure in the dark. Scientists have also discovered single-celled life forms that get their energy from chemicals rather than sunlight. Other scientists have found life surviving in hot springs that had been thought to be too hot to support life. Perhaps life forms exist that do not need the "Goldilocks conditions."

In 1970, a spacecraft found regions on the surface of Mars that look like streambeds with crisscrossing paths of water. **Since life as we know it requires water, scientists hypothesize that Mars may have once had the conditions needed for life to exist.** In 1976, twin *Viking* spacecraft each carried a small biology laboratory that tested to see if there were life forms on Mars that used oxygen and gave off carbon dioxide. None of these tests showed any evidence of life.

In 1996, scientists studied a meteorite from Mars found in Antarctica. They found tiny shapes that looked like fossils—remains of ancient life preserved in rock. Other scientists think that the shapes are not fossils but may have come from lumps of hardened clay formed on Mars. Or they may have been formed when snow got into cracks in the meteorite after landing on Earth.

Many scientists think that one of Jupiter's moons, Europa, may have conditions for life to develop. Photos from *Voyager* and *Galileo* showed that Europa has a very smooth icy crust with giant cracks in it. Could that mean that there is a liquid ocean under Europa's ice? The water in the ocean could possibly be kept warm by heat coming from inside Europa. **If there is liquid water on Europa, there might also be life.** People will have to wait for the next generation of spaces probes to find out.

# Is There Life Beyond Earth?

## ◆ Understanding Main Ideas

*Answer the following questions. Use a separate sheet if you need more room.*

**1.** What are three characteristics that all living things on Earth have in common?

_______________________________________________

**2.** What does the phrase "Goldilocks conditions" refer to?

_______________________________________________

**3.** What are the three "Goldilocks conditions"?

_______________________________________________

4. Scientist have discovered unusual life forms on Earth such as animals that live in the ocean at very high pressure in the dark and others that get their energy from chemicals. Using this information, what do scientists infer about the conditions necessary for life on other planets?

_______________________________________________

**5.** Spacecraft sent to Mars have found regions on the surface of Mars that look like stream beds. Why do these regions lead scientists to hypothesize that there may have once been life on that planet?

_______________________________________________

**6.** Why do scientist think that Europa might have the conditions for life to develop?

_______________________________________________

## ◆ Building Vocabulary

*Write a definition for the following term on the lines below.*

**7.** extraterrestrial life

_______________________________________________



2

Name ______________________ Date __________ Class ______________

SECTION 2-6 ENRICH

# Planets for Human Settlement

No one knows whether there is life on other planets, but one fact is clear. Humans have traveled through space. Will it ever be possible for humans to travel to other solar systems? Most people might not want to make such a difficult, expensive trip unless it ended at an Earthlike planet on which humans could easily live. What would such a planet have to be like?

2

Humans need several things, such as water, oxygen, light, and comfortable temperatures to support life. Whether a planet can supply what we need to live depends on many factors. Let's look at two important ones:

- **A planet's rotation rate.** A planet with a slow rotation rate has a long day, which would be harmful to humans. It gets hot in the afternoon and cold at night. What if day and night were each a week long? Even on Earth, we would roast and freeze in turns. Humans probably could not live comfortably on a planet whose day was more than four times as long as Earth's.
- **A planet's mass.** This determines many of its other characteristics. Larger planets have stronger gravity. Humans cannot stand gravity greater then 1.5 times Earth's for very long. Also, a planet's surface water and atmosphere originally come from beneath its surface. For this reason, a planet must be massive enough to have a large volume compared to its surface area. A planet smaller than 0.4 times Earth's mass would probably not have enough air or water to support human life.

The following table gives information about the mass, gravity, and day length of Earth, Venus, Mars, and three imaginary planets.

| | Earth | Venus | Mars | Planet A | Planet B | Planet C |
|---|---|---|---|---|---|---|
| Mass (Earth mass = 1) | 1 | 0.815 | 0.107 | 2.5 | 0.7 | 1 |
| Gravity (Earth gravity = 1) | 1 | 0.903 | 0.38 | 1.65 | 0.87 | 1 |
| Length of day (hours) | 24 | 5832.24 | 24.62 | 30 | 60 | 255 |

*Answer the questions below on the back of this sheet or on a separate sheet of paper.*

1. Planets A, B, and C revolve around stars exactly like the sun in orbits just like that of Earth. Could planet A possibly support human life? Explain.
2. Could planet B possibly support human life? Explain.
3. Could planet C possibly support human life? Explain.
4. If it were located in Earth's orbit, could Venus support human life? Explain.
5. If it were located in Earth's orbit, could Mars support human life? Explain.

# Stormy Sunspots

During which years were electrical disturbances on Earth most common? In this lab, you will consider the relationship between sunspot activity and magnetic storms on Earth.

◆ **Problem** How are magnetic storms on Earth related to sunspot activity?

◆ **Skills Focus** graphing, interpreting data

## ◆ Materials

graph paper pencil straightedge

## ◆ Procedure

1. Use the data in the table to plot a line graph of sunspot activity between 1967 and 1997.
2. On the graph, label the $x$-axis "year." Use a scale with 2-year intervals, from 1967 to 1997.
3. Label the $y$-axis "Average Number of Sunspots." Use a scale of 0 through 160 in intervals of 10.
4. Graph a point for the average number of sunspots for each year.

**Sunspots**

| Year | Average Number of Sunspots | Year | Average Number of Sunspots |
|---|---|---|---|
| 1967 | 93.8 | 1983 | 66.6 |
| 1969 | 105.0 | 1985 | 17.9 |
| 1971 | 66.6 | 1987 | 29.4 |
| 1973 | 38.0 | 1989 | 157.6 |
| 1975 | 15.5 | 1991 | 145.7 |
| 1977 | 27.5 | 1993 | 54.6 |
| 1979 | 155.4 | 1995 | 17.5 |
| 1981 | 140.4 | 1997 | 23.4 |

**REAL-WORLD LAB** *(continued)*

## ◆ Line Graph

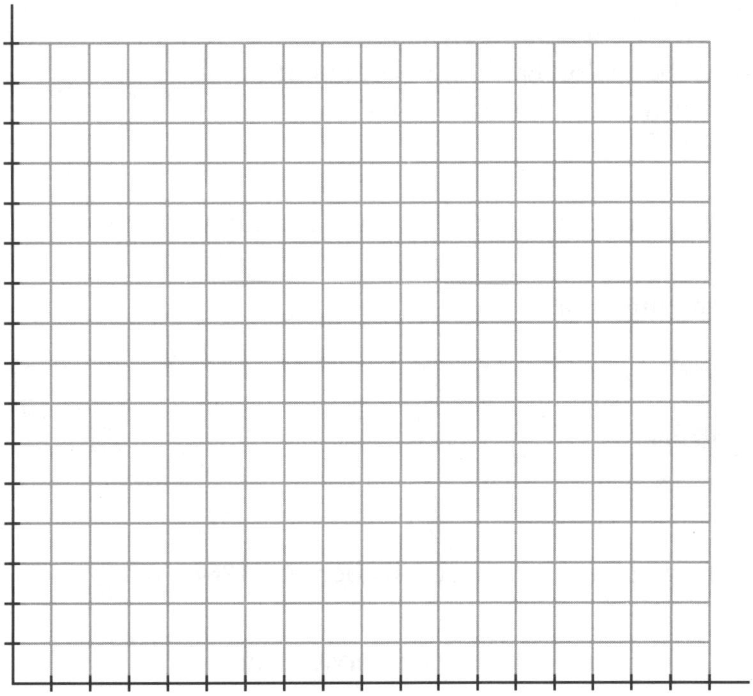

## ◆ Analyze and Conclude

*Write your answers on the back of this sheet or on a separate sheet of paper.*

1. Based on your graph, which years had the highest average number of sunspots? The lowest average number of sunspots?
2. How often does the cycle of maximum and minimum activity repeat?
3. When was the latest maximum sunspot activity? The latest minimum sunspot activity?
4. Compare your sunspot graph with the magnetic storms graph. What relationship can you infer between periods of high sunspot activity and magnetic storms? Explain.
5. **Apply** During which years do you think electrical disturbances on Earth were most common?

## ◆ More to Explore

Using the pattern of sunspot activity you found, predict the number of peaks you would expect in the next 30 years. Around which years would you expect the peaks to occur?

***Developing Hypotheses***

# Speeding Around The Sun

In this lab, you will make and test a hypothesis about how a planet's distance from the sun is related to its period of revolution.

◆ **Problem** How does a planet's distance from the sun affect its period of revolution?

## ◆ Materials

string, 1.5 m
plastic tube, 6 cm
weight or several washers
one-hole rubber stopper
stopwatch

## ◆ Procedure

*Review the safety guidelines in Appendix A.*

1. What do you think is the relationship between a planet's distance from the sun an its period of revolution? Write your hypothesis in the form of an "if . . . then" statement.
2. To test your hypothesis, you need to make a model planet.
   - **a.** Thread the string through the rubber stopper hole. Tie the end of the string to the main part of the string. Pull tightly to make sure that the knot will not become untied.
   - **b.** Thread the other end of the string through the plastic tube and tie a weight to that end. Have your teacher check both knots.
   - **c.** Hold the plastic tube in your other hand above your head. Practice keeping the stopper moving at a constant speed. The circle represents the planet's orbit. **CAUTION:** *Stand away from other students. Make sure the swinging stopper will not hit students or objects. Do not let go of the string.*
3. Before you try different distances for your model planet, look at the data table on the next page.
4. Pull the string so the stopper is 20 cm away from the plastic tube. Swing the stopper just fast enough to keep the stopper moving.

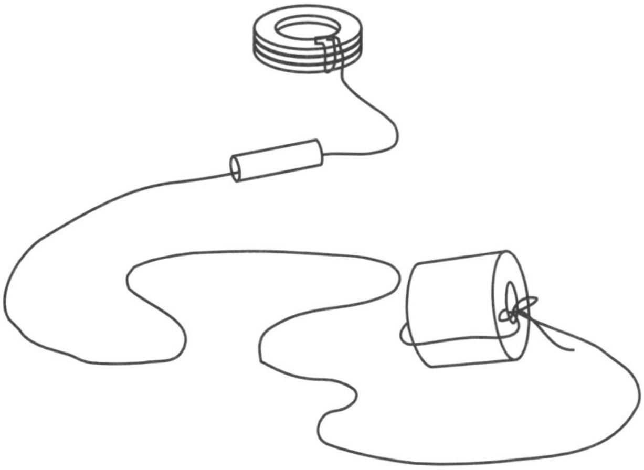

## SKILLS LAB *(continued)*

5. Have your partner time how long it takes for the stopper to make 10 revolutions. Divide by 10 to find the period of revolution. Record this number as Trial 1.
6. Repeat Steps 4–5 two more times. Record your results as Trials 2 and 3. Add the results of the three trials together and divide by three to find the average period of revolution.
7. If you pull the stopper out to 40 cm, do you think the period of revolution will increase or decrease? To find out, pull the stopper out to 40 cm and repeat Steps 4–6.



8. Based on your results in Step 7, do you want to revise your hypothesis? Make any needed changes, then pull the stopper out to 60 cm and repeat 4–6.

### ◆ Data Table

| Distance (cm) | Period of Revolution | | | |
|---|---|---|---|---|
| | Trial 1 | Trial 2 | Trial 3 | Average |
| 20 | | | | |
| 40 | | | | |
| 60 | | | | |

Name ______________________ Date __________ Class ______________

SKILLS LAB *(continued)*

## ◆ Analyze and Conclude

*Write your answers on the back of this sheet or on a separate sheet of paper.*

1. Which object in your model represented the sun? Which represented the planet?
2. What force did the pull on the string represent?
3. When you pulled the stopper out to make the orbit larger, did the string then represent a stronger or weaker force of gravity? Why?
4. What happened to the period of revolution when you made the orbit larger in Steps 7 and 8?
5. Did your observations support your hypothesis? Summarize your conclusions based on your observations.
6. Which planets take less time to revolve around the sun—those closer to the sun or those farther away? Use the model to support your answer.
7. **Think About It** What information did you consider when you made your hypothesis? How did having some experimental data help you modify your hypothesis?

## ◆ Design an Experiment

Write a hypothesis relating the mass of a planet to its period of revolution. Then, using a stopper with a different mass, modify the activity to test your hypothesis. Before you swing the stopper, have your teacher check your knots.

2

CHAPTER 3

# Star Stories

The following steps will walk you through the Chapter 3 Project. Use the hints as you guide students through choosing a constellation, researching it, developing a new myth, and presenting their new myth to the class.

- This project asks students to bring together a number of different skills: using star charts to locate constellations, researching cultural history and background, thinking creatively, and communicating in writing. The goal is for students to research the stories that various cultures have told about a given constellation. Then they will make up their own name for the pattern they see in that constellation and write a story to support it.
- In this project, students get acquainted with classical mythology and also discover ways in which non-European cultures have interpreted the patterns of stars. They use their imaginations to write their own stories, following the steps of the writing process outlined in Worksheet 2 on page 75.
- Have students read the Chapter 3 Project Overview, pages 72–73. Review the project's rules and hand out the Chapter 3 Scoring Rubric, page 76, that you will use in scoring students' work. Discuss with students what will be expected of them.
- Set a deadline for the paper submission and project presentation and interim dates for the Check Your Progress at the end of Section 1, Section 3, and Section 5. Encourage students to copy the dates in the Project Time Line.
- Have students choose three constellations to try to observe in the night sky using the star charts in Appendix B. Make sure that at least one of the constellations will be easy to see as this will build their confidence. Check the night sky a few days before beginning the project to determine which constellations are easiest to see.
- Distribute copies of Chapter 3 Project Worksheets 1 and 2 on pages 74–75. Worksheet 1 will help students to get involved in the project and Worksheet 2 will help them write their story. Remind students to refer to the hints in the Chapter 3 Project Overview as they plan and carry out the project.

## ◆ Materials and Preparation

- Gather research resources together for the students. These might include:

  Milton Heifetz and Wil Tirion. *A Walk Through the Heavens: A Guide to Stars and Constellations and Their Legends.* Cambridge University Press, 1996. (Note: Much of the violence in the mythology has been toned down.)

  Chet Raymo. *365 Starry Nights: An Introduction to Astronomy for Every Night of the Year.* Prentice Hall Press, 1982.

  John Gustafson. *Stars, Clusters and Galaxies. The Young Stargazer's Guide to the Galaxy.* Julian Messner, 1992.

  Fred Schaaf. *Seeing the Sky. 100 Projects, Activities and Explorations in Astronomy.* John Wiley and Sons, 1990. (Note: The projects are generally for adults.)

## ◆ Check Your Progress—Section 1 Review

- Guide students' observations by pointing out constellations that are visible in the early evening (if cloudy weather makes observations impossible, students can study constellations from the star charts alone).
- Reinforce the point of the Section 1 Discover activity that the stars in constellations are not really neighbors in space. (The Pleiades are an exception.) Also, students should understand

**CHAPTER 3 PROJECT** *(continued)*

that people in many different parts of the world have identified the same or similar patterns of stars in the sky.

- Provide class time for students to tell some of the myths they find. From the resources suggested, you can give examples of how different cultures have interpreted the same constellation differently. You may want to have them discuss why, for example, farming peoples might associate certain stars with annual floods or seasonal changes. These stories and legends set the stage for the next step when students will write their own stories to explain a star pattern.
- Be sure to explain that one aspect of myths is their variability. They were passed down from person to person, and many versions of the same myths exist. The students aren't wrong if their stories differ from other students'.

## ◆ Check Your Progress—Section 3 Review

- These writing exercises are the first step in the writing process, called prewriting. Along with researching, prewriting techniques can include brainstorming ideas, listing possible subjects, narrowing the focus, identifying the audience, and making an outline.
- If students are having trouble thinking of ideas for re-imagining and renaming their constellations, have them brainstorm with several classmates. Students may also need help focusing on one pattern and name.

## ◆ Check Your Progress—Section 5 Review

- The middle steps are perhaps the most important in the writing process. Be sure that students actually go through the two distinct steps of writing a first draft and then revising/editing it. Their stories do not have to be longer than an anecdote of one or two paragraphs, although some students will write more elaborate explanations. Remind students that their final goal is to re-imagine the pattern of stars in a constellation, name it, and then write a story that explains the name they have chosen.

## ◆ Chapter Project Wrap-Up

- Students' final presentation generally will show whether they have done the appropriate research and completed the writing assignment. As you discuss with students the form of their final presentation, you will be able to assess the quality of their research. Students should show familiarity with at least the myths associated with their constellation. One likely choice of presentation is a poster illustrating the constellation and the story that students have invented, but be receptive to other ideas. Encourage students who have studied the same constellation to discuss their different approaches to it.

## ◆ Extensions/Options

- Although it is important for students to locate and recognize constellations in the night sky, you can get more variety in research projects by letting students choose constellations that are not visible at this season. To ensure variety and avoid too many students choosing the same constellation, you may want to assign students the constellations that they will research.

CHAPTER 3

# Star Stories

For this project, you will first use a star chart from Appendix B of your textbook to identify three constellations. Using your star chart, you will go outside and identify these constellations in the night sky. You will then research the name of one constellation, finding stories about the constellation from different cultures. After the research, you will rename the constellation based on what the star pattern looks like to you. You will then come up with a story to explain the new name of the constellation. Finally, you will write your story and present it to the rest of the class.

3

## ◆ Project Rules

- Have your teacher approve the constellations that you choose.
- Find as many stories about your constellation as you can.
- After writing a rough draft, show it to your teacher to make sure that you are on the right track.

## ◆ Suggested Materials

- You will need research sources, such as encyclopedias and books on mythology. Worksheet 1 summarizes the myths of several cultures that apply to some major constellations.
- You may need art materials if you choose to make posters for your final presentation. These might include poster board, glue, drawing paper, pens, and pencils.

## ◆ Project Hints

- Carefully study the star maps in Appendix B in the textbook before looking for constellations in the night sky. Try to find an area away from outside lights to view constellations.
- Libraries and the Internet are good places to find information about the constellation that you choose to research. Worksheet 1 may also be helpful in getting you started.
- When trying to decide upon a new name for the constellation that you have chosen, sketch the positions of the stars in the constellation, and then spend some time looking at your sketch from many different angles. Write down all the things that you see in your constellation.
- When you are trying to come up with a story for your constellation, do not be afraid to use your imagination. There is no "wrong" story.
- Before you begin writing your story, go over Worksheet 2, which gives an outline of the stages of writing that you should follow.

3

## ◆ Project Time Line

| Task | Due Date | Teacher's Initials |
|---|---|---|
| **1.** Three constellations selected and observed | ________ | ________ |
| **2.** One of these three constellations chosen for research | ________ | |
| **3.** Research of story behind chosen constellation completed | ________ | |
| **4.** New story formulated for constellation | ________ | |
| **5.** First written draft of story shown to teacher | ________ | ________ |
| **6.** Final written draft turned in and story presented to class | ________ | |

Name ______________________ Date __________ Class ______________

CHAPTER 3 PROJECT WORKSHEET 1

# Star Stories

Orion: Orion the Hunter is one of the easiest constellations to see in the winter sky, especially the three bright stars that make up his belt. In Greek myth, Orion was a mighty hunter, and he boasted that he could wipe out all the animals on Earth. As punishment for bragging, the Earth goddess Gaia sent a poisonous scorpion to bite him. (The scorpion is also a constellation, called Scorpio, which is at the opposite end of the sky, so the two will never meet again). Arabic cultures and Indians in Brazil also viewed the Orion constellation as a hunter.

Two hunting dogs follow Orion—the Great Dog, Canis Major, near his foot and the Little Dog, Canis Minor, at his shoulder. Sirius, the "Dog Star," is at the head of the Great Dog. It is the brightest star in the night sky.

3

The Egyptians saw the constellation Orion as the major god Osiris, who dies and is reborn each year as the constellation disappears for 70 days and then reappears in the sky.

Orion's belt had many names, too. European sailors called it the "Yardarm," which is the name for the crosspiece of the mast on a sailing ship. In parts of Latin America, the three stars are the "Three Kings."

### Stars, Constellations and Their Myths

| Constellation/Star | Classical Story (Greek/Roman) | Other Stories |
|---|---|---|
| Pleiades | Seven sisters or flock of doves | Native American: Lost children |
| Lyra | Orpheus' lyre or harp. Romans saw a vase. | Persia: tortoise or clay tablet. Also, eagle, falcon, or vulture. |
| Northern Crown (Corona Borealis) | Crown of jewels put in sky by Bacchus as a wedding present for Ariadne | Native American: Council of chiefs in a circle, or den of the Great Bear.<br>Arabia: "Beggar's Bowl"<br>Siberia: "Polar Bear's Paw"<br>China: coins on a string |
| Virgo | Identified with goddess of harvest (Persephone), holding sheaf of wheat, or as goddess of justice | Ancient Britons: harvest goddess Eostre.<br>Ancient Egypt: identified with goddess Isis. |
| Sirius (Dog Star) | Name means "scorching." Romans blamed it for hot weather. | Ancient Egypt: Sothis or Anubis, the jackal god. Believed to cause the yearly Nile floods. |

Name ______________________ Date __________ Class ______________

CHAPTER 3 PROJECT WORKSHEET 2

# Star Stories

*When writing a paper, it is useful to go through the following sequence:*

1. **Prewriting:** During the prewriting stage, you will perform any activities that must be done before the actual writing of your story. These include:
   - ◆ researching your constellation to become familiar with the stories told by various cultures about it
   - ◆ brainstorming ideas for a possible story of your own concerning the constellation
   - ◆ narrowing the focus so that you decide upon a particular story to write about
   - ◆ identifying the audience that you will be writing the story for (your teacher and classmates)
   - ◆ making an outline of the story and how you will write about it
2. **Writing a first draft:** You need to write a first draft so that you have something that you can edit and revise as needed. Writing a first draft, then revising and editing the first draft are perhaps the most crucial steps in the writing process.
3. **Revising the first draft:** Once you have the first draft, you will need to revise and edit it. You might want to add particular details to the story that you overlooked in the first draft. You might wish to change the introduction or the ending, revise sentences to clarify them, change words or sentences to make your writing more interesting, and choose more exact words. The content of the first draft, how it is presented, and the language used should be carefully revised.
4. **Writing a final draft:** Rewrite your story, incorporating changes you made during the revising stage. Then proofread the final version. Carefully check your spelling, vocabulary, grammar, capitalization, and punctuation.
5. **Publishing:** You then have to decide how you will present the final draft of your story, whether in manuscript form, in an audiovisual presentation, or in some other form.

3

CHAPTER 3

# Star Stories

In evaluating how well you complete the Chapter 3 Project, your teacher will judge your work in four categories. In each, a score of 4 is the best rating.

| | 4 | 3 | 2 | 1 |
|---|---|---|---|---|
| ***Research into Constellation*** | Thoroughly researches the myths involving the chosen constellation. Able to tell the original myth and answer questions about the myth. | Adequately researches the myths behind the chosen constellation. Able to tell the original myth in outline and answer some questions about the myth. | Has done some research on the myths. Unable to recount many of the details of the myth. | Has done very little research on the myths. Has little familiarity with the myths. |
| ***First Draft of Paper*** | First draft finished and well prepared by due date. | First draft finished but poorly prepared on due date. | First draft unfinished by due date. | First draft barely started by due date. |
| ***Final Draft of Paper*** | Paper contains a complete, easy to follow story of the new constellation name. Paper is well-written, with no spelling or grammatical errors. | Paper contains a complete story of the new constellation name. Paper has a few spelling or grammatical errors. | Paper contains a somewhat incomplete story of the new constellation name. Paper is poorly written, with spelling or grammatical errors. | Paper contains an incomplete story of the new constellation name. Paper is poorly written, with many spelling and grammatical errors. |
| ***Presentation*** | Identifies constellation and clearly and creatively presents new name and story behind the name. | Identifies constellation. New name and story behind the name are adequately presented. | Identifies constellation. New name and story behind the name are presented, but not clearly. | Does not adequately identify constellation, and does not adequately present story behind new name. |

*Integrating Physics*

# Tools of Modern Astronomy (pp. 94–102)

## Objectives

After completing the lesson, students will be able to

- describe the electromagnetic spectrum;
- explain the main purpose of a telescope;
- explain why astronomers use spectrographs.

## 1 Engage/Explore

- **Activating Prior Knowledge** Encourage students to describe the stars they have seen in the night sky.
- **Discover** After modeling a constellation, have students look at it with one eye and challenge them to determine which balls are farther away and which are closer.

## 2 Facilitate

- **Using the Visuals: Figure 1** Ask students whether red or yellow light has a longer wavelength.
- **Building Inquiry Skills: Observing** Allow students to observe the effects of a prism on white light.
- **Try This** Invite students to determine in which direction a radio station is located using an umbrella to focus radio waves.
- **Including All Students** Help students distinguish refracting and reflecting telescopes by having them sketch each telescope. Then have students explain how each telescope works.
- **Using the Visuals: Figure 3** Guide students to discover how radio telescopes have enabled astronomers to expand their exploration of space.
- **Addressing Naive Conceptions** Make sure students understand that the stars are also in the sky during the day.
- **Science and History** Help students realize how large the telescopes are.
- **Integrating Technology** Challenge students to speculate on the advantages of placing telescopes in space instead of on Earth.
- **Inquiry Challenge** Encourage students to predict and observe what will happen when different colored cellophane is held between a prism and a spectrum projected from the prism.
- **Sharpen Your Skills** Have students infer which elements have the strongest lines in each star spectrum.

## 3 Assess

- **Section 1 Review** Check students' understanding of section content by having them answer the section review questions.
- **Performance Assessment** Assess students' understanding of telescopes and spectrographs by having them write paragraphs describing some of the tools and procedures they would use to investigate a new star.

## Lab

- **Real World Lab: How It Works** Allow students to construct and use a simple refracting telescope.

## Science and Society

- **Light Pollution** After explaining the problems of light pollution, direct students to list possible solutions. Make sure they list the procedures that are involved in each solution and advantages and disadvantages of each solution.

3

Name ______________________ Date __________ Class ______________

SECTION 3-1 SECTION SUMMARY

# Tools of Modern Astronomy

**Guide for Reading**

- What is the electromagnetic spectrum?
- What is the main purpose of a telescope?
- Why do astronomers use spectrographs?

3

Patterns of stars in the sky are called **constellations.** Stars in constellations may look as if they are close together, but they are actually very far apart. They just happen to be in the same direction in the sky.

Scientists call the light you see with your eyes **visible light.** Light is a form of **electromagnetic radiation,** or energy that can travel directly through space in the form of waves. The distance between the crest of one wave and the crest of the next wave is called the **wavelength.** If you shine white light through a prism, the light separates out to make a range of different colors with different wavelengths, called a **spectrum.** The spectrum of visible light is composed of the colors red, orange, yellow, green, blue, and violet. **The electromagnetic spectrum includes radio waves, infrared radiation, visible light, ultraviolet radiation, X-rays, and gamma rays.**

Modern astronomy is based on the detection of electromagnetic radiation from objects in space. **Most telescopes collect and focus different types of electromagnetic radiation, including visible light.**

A **refracting telescope** uses convex lenses to gather a large amount of light and focus it onto a smaller area. A **convex lens** is a piece of transparent glass, curved so that the middle is thicker than the edges. Refracting telescopes use two lenses—an eyepiece lens and an objective lens. When light passes through the objective lens, the lens focuses the light at a certain distance away from the lens. This distance is called the focal length of the lens. Isaac Newton built the first **reflecting telescope.** This telescope collects and focuses light using a mirror. Devices used to detect radio waves from space are called **radio telescopes.** Other telescopes detect ultraviolet, infrared, X-ray, and gamma radiation.

A building that contains one or more telescopes is called an **observatory.** Because Earth's atmosphere makes objects in space look blurry, observatories are often built on mountaintops. To gather information on types of radiation blocked by Earth's atmosphere, scientists have placed telescopes on satellites.

A **spectrograph** breaks the light from an object into colors and photographs the resulting spectrum. **Astronomers use spectrographs to get information about stars, including their chemical composition and temperatures.** Each chemical element absorbs light at different wavelengths. These appear as dark lines in a spectrum. The amount of radiation absorbed by an element depends on its temperature. Astronomers can find out the star's temperature by comparing a star's spectrum with the spectrums of elements at different temperatures.

Name ______________________ Date __________ Class ______________

**SECTION 3-1** **REVIEW AND REINFORCE**

# Tools of Modern Astronomy

## ◆ Understanding Main Ideas

*For each tool on the left, draw a line to connect it with its function. Then continue the same line to connect the tool and function with the type of electromagnetic radiation that the tool uses. Some functions and types of electromagnetic radiation will have more than one line connecting them. The first one is done for you.*

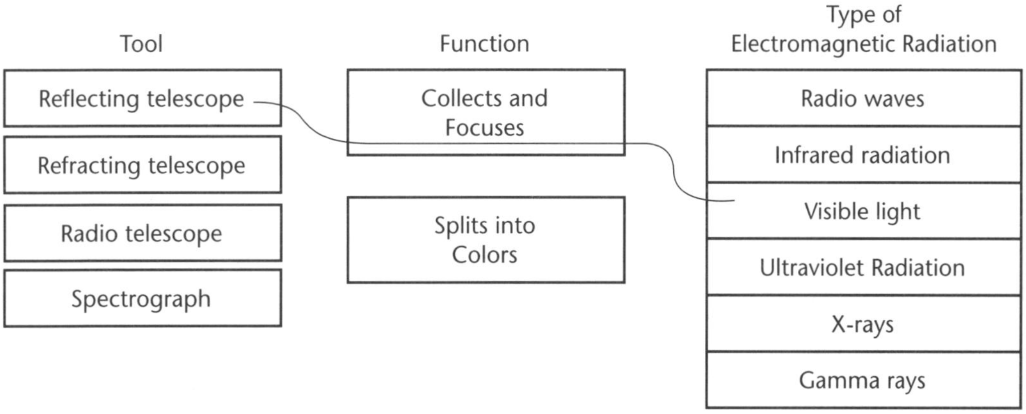


**1.** Name two things astronomers can learn by studying the spectrums of stars.

________________________________________________________________

## ◆ Building Vocabulary

*Match each term with its definition by writing the letter of the correct definition in the right column on the line beside the term in the left column.*

| | |
|---|---|
| ________ **2.** constellation | **a.** a building that contains one or more telescopes |
| ________ **3.** spectrum | **b.** the portion of the electromagnetic spectrum that humans can see with their eyes |
| ________ **4.** observatory | **c.** white light shining through a prism spreads out to form one |
| ________ **5.** wavelength | **d.** a pattern of stars in the sky |
| ________ **6.** visible light | **e.** an instrument that breaks the light from an object into colors and photographs the resulting spectrum |
| ________ **7.** spectrograph | **f.** the distance between the crest of one wave and the crest of the next wave |

3

Name ______________________ Date __________ Class ______________

SECTION 3-1 ENRICH

# Seeing Without Light

Some of the most interesting objects in the sky, such as new stars, are hidden behind the huge clouds of gas and dust that lie among the stars. Astronomers now "see" through these barriers using parts of the electromagnetic spectrum that have wavelengths longer than visible light.

An astronomer deciding which part of the electromagnetic spectrum to use must deal with at least two problems:

- **Some wavelengths don't pass through the atmosphere:** Earth's atmosphere allows visible light, radio, and some infrared waves to pass through, but stops ultraviolet, X-ray, and gamma rays. Astronomers who want to use wavelengths other than visible light, infrared, and radio waves must make their observations from satellites.
- **Some wavelengths produce images with low resolution:** *Resolution* is the ability to distinguish objects that are close together. Low-resolution images are fuzzy. High-resolution images are sharp. Generally, the larger the diameter of the telescope's collecting surface, the higher the resolution of the image produced. And the shorter the wavelength of radiation used, the higher the resolution of the image that can be produced. Radio telescopes, despite their large diameter, produce very low resolution images because of the long wavelengths of radio waves. Infrared, on the other hand, has wavelengths closer to visible light.

3

*Answer the questions below on the back of this sheet or on a separate sheet of paper.*

| Radiation Type | Passes Through Atmosphere? | Wavelength |
|---|---|---|
| Radio | Yes | >1 mm |
| Infrared | Some | 0.0007 mm–1 mm |
| Visible light | Yes | 0.0004 mm–0.0007 mm |
| Ultraviolet | No | 0.000 02 mm–0.0004 mm |
| X-ray | No | 0.000 000 01 mm–0.000 02 mm |
| Gamma | No | <0.000 000 1 mm |

1. Which form of radiation should give the highest resolution image? Why?
2. What do you suppose are the disadvantages of satellite-based observations?
3. Suppose astronomers are interested in obtaining an image of a very large area in the sky that gives off equal amounts of radio waves, infrared waves, and visible light. Which radiation should they observe if they want to get images 24 hours a day, quickly and cheaply? Why?
4. If astronomers wanted a very detailed image of an object behind a dust cloud, which radiation should they observe? Why?

# Characteristics of Stars (pp. 103–111)

## Objectives

After completing the lesson, students will be able to

- describe how astronomers measure distances to nearby stars;
- explain how stars are classified.

## 1 Engage/Explore

- **Activating Prior Knowledge** After displaying a star chart showing the current night sky, invite students to discuss how they could use star charts to locate objects in the sky.
- **Discover** Invite students observe how far their thumb moves and how far away it is when they look at it with the right eye and then the left.

## 2 Facilitate

- **Building Inquiry Skills: Making Models** To help students remember that a light-year measures distance, not time, have them calculate the distance light travels in one minute, one day, and one hour.
- **Inquiry Challenge** Challenge students to measure the distance to an object using triangulation.
- **Using the Visuals: Figure 6** Ask students to describe the difference between the images of the January and July skies.
- **Building Inquiry Skills: Classifying** Have students observe stars and record the different colors they see.
- **Using the Visuals: Figure 7** Remind students that the differences in the sizes of stars is actually much greater than shown.
- **Building Inquiry Skills: Making Models** To help students understand the range of star sizes, allow students to model each type of star to scale.
- **Using the Visuals: Figure 8** Have students look at the figure and make inferences about the relative temperatures of each star according to its color.
- **Including All Students** Invite students to think of words other than *photosphere* that contain the word *photo.*
- **Social Studies Connection** Using examples, inform students that the Middle Ages coincided with a very active time in Arabic sciences.
- **Try This** Have students model the apparent magnitude of stars using flashlights which vary in brightness.
- **Including All Students** Have partners take turns asking and answering questions using the Hertzsprung-Russell diagram.

## 3 Assess

- **Section 2 Review** Check students' understanding of characteristics of stars by having them answer the section review questions.
- **Performance Assessment** Assess students' understanding of the characteristics of stars by having them prepare a presentation in which they define the characteristics of a new star, compare it to our sun, and show the placement of their star on the Hertzsprung-Russell diagram.

## Lab

- **Skills Lab: Measuring** Challenge students to examine how parallax can be used to determine distances of stars by setting up a model of a telescope.

Name ______________________ Date __________ Class ______________

SECTION 3-2 SECTION SUMMARY

# Characteristics of Stars

**Guide for Reading**

- How do astronomers measure distances to nearby stars?
- How are stars classified?

3

The sun and nearby stars are part of the Milky Way. The Milky Way is a giant flat structure called a **galaxy,** which contains hundreds of billions of stars. There are billions of galaxies in the **universe,** which astronomers define as all of space and everything in it. Astronomers use a unit called a **light-year** to measure distances in space. A light-year is the distance that light travels in one year, or about 9.5 million million kilometers. Note that the light-year is a unit of distance, not time.

**Astronomers use parallax to measure distances to nearby stars. Parallax** is the apparent change in position of an object when you look at it from different places. To measure parallax, astronomers look at a star when Earth is on one side of the sun. Then they look at the same star again six months later, when Earth is on the other side of the sun. They then measure how much the star appears to move against the background of stars that are much farther away. The closer the star is to Earth, the more it appears to move. They can then use this measurement, called the parallax shift, to calculate how far away the star is.

**The main characteristics used to classify stars are size, temperature, and brightness.** Many stars are about the size of the sun, which is a medium-sized star. Very large stars are called **giant stars** or supergiant stars. Some stars are smaller than the sun. White dwarf stars are about the size of Earth. Neutron stars are even smaller, only about 20 kilometers across.

You can tell a star's temperature by its color. The coolest stars glow reddish and are about 3,500 degrees Celsius. The sun glows white and has a surface temperature of 6,000 degrees Celsius. The hottest stars are slightly bluer than the sun and have surface temperatures over 10,000 degrees Celsius.

The brightness of a star can be described by **apparent magnitude** and **absolute magnitude.** A star's apparent magnitude is its brightness as seen from Earth. A star's absolute magnitude is the brightness the star would have if it were at a standard distance from Earth. To find a star's absolute magnitude, an astronomer must know a star's apparent magnitude and how far it is from Earth.

Two important characteristics of stars are temperature and absolute magnitude. Ejnar Hertzsprung and Henry Norris Russell made graphs to find out if temperature and brightness are related. The graph they made is called the **Hertzsprung-Russell diagram,** or H-R diagram. Most of the stars in the H-R diagram form a diagonal line called the **main sequence.** In the main sequence, surface temperature increases as brightness increases.

Name ______________________ Date __________ Class ______________

SECTION 3-2 REVIEW AND REINFORCE

# Characteristics of Stars

## ◆ Understanding Main Ideas

*Use the following H-R diagram to answer questions 1–3. Write your answers in the spaces provided.*

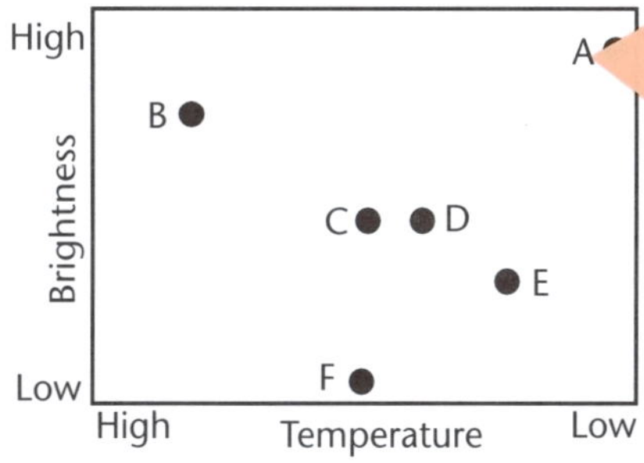


________ **1.** Which star has the greatest brightness?

________ **2.** Which has the lowest brightness?

________ **3.** Stars C and D have the same brightness. What would you need to know to determine their relative magnitudes?

*Answer the following questions on a separate sheet of paper.*

**4.** Explain how astronomers measure the distance to nearby stars.

**5.** What are the main characteristics used to classify stars?

**6.** How would you classify the sun based on each of these characteristics?

## ◆ Building Vocabulary

*From the list below, choose the term that best completes each sentence.*

| | | |
|---|---|---|
| galaxy | universe | light-year |
| parallax | giant star | apparent magnitude |
| absolute magnitude | Hertzsprung-Russell diagram | main sequence |

**7.** A star's brightness as if it were a standard distance is its ______________.

**8.** A huge group of stars, such as the Milky Way, is a(n) ______________.

**9.** The unit used to measure distances between stars is a(n) ______________.

**10.** The region of the Hertzsprung-Russell diagram with the most stars is the ______________.

**11.** A graph of stars showing temperature on the *x*-axis and brightness on the *y*-axis is a(n) ______________.

**12.** ______________ is used to determine the distance to nearby stars.

**13.** The ______________ is all of space and everything in it.

**14.** The brightness of a star as seen from Earth is its ______________.

**15.** Because of its large size, the star Aldebaran is often called a(n) ______________.

# The Apparent Magnitude Scale

The apparent magnitude of stars was first recorded by the Greek astronomer Hipparchus about 160 B.C. Hipparchus grouped stars according to their brightness, or *magnitude.* He called the twenty brightest stars *first magnitude* stars. Stars half that bright were *second magnitude. Third magnitude* stars were half as bright as second magnitude stars, and so on.

Modern astronomers have changed Hipparchus's system a little. In the modern system, an object of magnitude 1 is 100 times as bright as one of 6. This means that if two objects are separated by one unit of magnitude, the brighter object is about 2.5 times as bright as the dimmer one. A star of magnitude 1 is about 2.5 times as bright as a star of magnitude 2. Using this scale, the *dimmer* the object, the *larger* the magnitude number.

3

What about very bright objects? Remember, the brighter an object appears, the lower its magnitude number. Very bright objects have negative magnitudes. Sirius, the brightest star in the sky (except for the sun), has a magnitude of −1.5. Venus is the brightest planet. Its brightness varies, but at its brightest, its magnitude is −4. The sun's magnitude is −26.5.

| Object | Apparent Magnitude |
|---|---|
| Sun | −26.50 |
| Venus (maximum brightness) | −4.00 |
| Mars (maximum brightness) | −2.00 |
| Hadar | 0.63 |
| Procyon | 0.36 |
| Regulus | 1.36 |
| Dimmest visible to the naked eye | 6.00 |
| Dimmest visible with binoculars | 10.00 |

*Refer to the table above to answer the questions below on the back of this sheet or on a separate sheet of paper.*

1. What is the difference between the magnitudes of the bright stars Regulus and Procyon? Which is brighter? How much brighter is it than the dimmer one?
2. How much brighter is Venus than Mars?
3. What is the difference between the magnitude of the dimmest star that is visible with the naked eye, and the dimmest that can be seen with binoculars? How much brighter is the former than the latter?

# Lives of Stars (pp. 112–116)

## Objectives

After completing the lesson, students will be able to

- explain how the life of a star begins;
- explain what determines how long a star will live;
- describe what happens to a star when it runs out of fuel.

## 1 Engage/Explore

- **Activating Prior Knowledge** Encourage students to describe the stages of a burning campfire. Lead students to see that wood is the fuel that feeds the flame.
- **Discover** Have students discover the general relationship between a star's mass and its lifetime.

## 2 Facilitate

- **Building Inquiry Skills: Problem Solving** Ask students to explain how they determine people's ages from pictures. Then ask them how they think astronomers can determine a star's age.
- **Using the Visuals: Figure 11** Have students explain the relationship between the object in the large image and the object in the small image. Make sure students realize that the Orion Nebula may contain hundreds of young stars.
- **Demonstration** Help students visualize how the more compact matter is, the more mass it has, by determining the mass of a beaker filled with marbles, then determining the mass of the same beaker filled with sand.
- **Sharpen Your Skills** Direct students to find Algol, Polaris, and Sirius B in the H-R diagram on page 109. For each star, tell them to write a sentence predicting what the next stages in its life will be.
- **Exploring the Lives of Stars** Have students volunteer to represent the stages of the life cycle of stars. Have them use this method to compare and contrast the different stages of giant stars and medium-sized stars.
- **Addressing Naive Conceptions** Inform students that the term *black hole* refers to the fact that the gravity of black holes is so intense that not even light can escape from them, not that they are black in color.

## 3 Assess

- **Section 3 Review** Check students' understanding of the lives of stars by having them answer the section review questions.
- **Performance Assessment** Assess students' understanding of the lives of stars by having them write short biographies detailing the life cycle of a medium-sized star.

SECTION 3-3 SECTION SUMMARY

# Lives of Stars

**Guide for Reading**

- How does the life of a star begin?
- What determines how long a star will live?
- What happens to a star when it runs out of fuel?

3

In 1967, Jocelyn Bell found an object in space that gave off regular pulses of radio waves. Stars like the one Bell discovered are called **pulsars,** short for pulsating radio sources.

Each star is born, goes through its life cycle, and dies. Astronomers can't watch a single star for billions of years so they study many stars in different stages of their lives and see how they differ from each other.

A **nebula** is a large amount of gas and dust spread out in an immense volume. Gravity can pull some of the gas and dust in a nebula together. The contracting cloud is then called a **protostar.** A protostar is the earliest stage of a star's life. **A star is born when the contracting gas and dust becomes so hot that nuclear fusion starts.**

**How long a star lives depends on how much mass it has.** Small stars use their fuel more slowly than large stars, so they have much longer lives.

When a star begins to run out of hydrogen fuel, the star becomes a red giant or supergiant. **When a star runs out of fuel, it becomes a white dwarf, a neutron star, or a black hole.**

When small or medium stars use up their fuel, their outer layers expand. At this stage they are called red giants. Eventually, the outer parts grow bigger and drift into space. The blue-white hot core of the star that is left behind is a **white dwarf.**

A dying giant or supergiant star can suddenly explode. The explosion is called a **supernova.** After the star explodes, some of the materials from the star are left behind. This forms a neutron star. **Neutron stars** are even smaller and denser than white dwarfs.

The most massive stars become **black holes** when they die. After a large star explodes, a large amount of mass may remain. The gravity of the mass is so strong that gas is pulled inward, pulling more gas into a smaller and smaller space. Eventually, the gravity becomes so strong that nothing can escape, not even light.

Astronomers have discovered several objects that were very bright, but also very far away. These distant bright objects looked almost like stars and were given the name **quas**i-stell**ar** objects, or **quasars.** Quasars are distant galaxies with giant black holes at their centers.

Name ______________________ Date __________ Class __________

**SECTION 3-3** **REVIEW AND REINFORCE**

# Lives of Stars

## ◆ Understanding Main Ideas

*Fill in each blank with the correct letter from the diagram.*

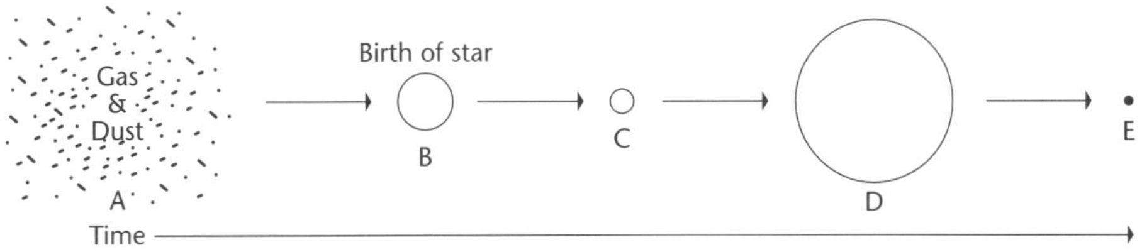


________ **1.** Red giant or supergiant

________ **2.** Where fusion begins

________ **3.** Part of a nebula

________ **4.** White dwarf, neutron star, or black hole

________ **5.** The stage the sun is in

*Fill in the blank.*

**6.** How long a star lives and what it becomes at the end of its life depend primarily on its ______________ .

## ◆ Building Vocabulary

*From the list below, choose the term that best matches each phrase.*

________ **7.** pulsar

________ **8.** white dwarf

________ **9.** nebula

________ **10.** protostar

________ **11.** supernova

________ **12.** neutron star

________ **13.** black hole

________ **14.** quasar

**a.** exerts such a strong gravitational pull that no electromagnetic radiation can escape

**b.** a huge cloud of gas or dust in space

**c.** what an average-sized star becomes at the end of its life

**d.** the earliest stage of a star's life

**e.** emits pulses of radio waves

**f.** shines brightly in the center of a distant galaxy because of the friction of material spiraling around it

**g.** formed from the leftover material after a giant star explodes, with three times the mass of the sun, but a diameter of only about 20 kilometers

**h.** an explosion that occurs at the end of a giant star's life

SECTION 3-3 ENRICH

# Stellar Remnants

Medium-sized stars, such as the sun, eventually consume all of the nuclear fuel available. The core shrinks and becomes hotter, while the surface expands. The star becomes a red giant. In the very last stage of its life as a star, the core collapses into a white dwarf, while the remaining hydrogen drifts away into a shell of gas called a planetary nebula. Planetary nebulas are considered by many astronomers and photographers to be the most beautiful astronomical objects that can be seen with a telescope. They are called planetary nebulas because the astronomers who first saw them thought they resembled planets in size and color. One of the most visually attractive planetary nebulas is found in the constellation Lyra. It is called the Ring nebula.

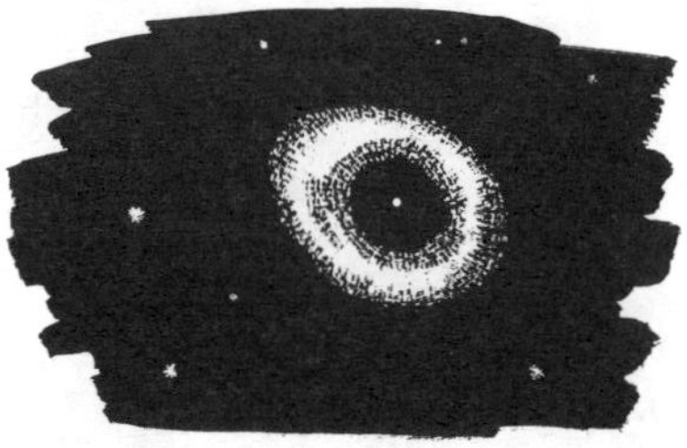
Figure 1

3

The gas surrounds the white dwarf in a three-dimensional sphere. Why does a three-dimensional spherical shell of gas appear to be a two-dimensional ring? Figure 2 shows the cross section of a shell of gas around a white dwarf. If you look at the shell along line 1 or 3, it appears bright. But if you look at the shell along line 2, it is less bright.

Figure 2

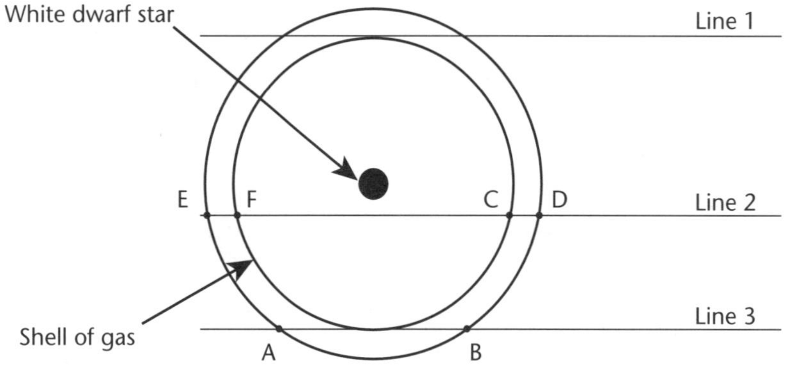


*Examine the figures and answer the questions below on the back of this sheet or on a separate sheet of paper.*

**1.** How far is it from A to B in millimeters?

**2.** How far is it from C to D in millimeters, plus from E to F?

**3.** Are you looking through more of the shell along line 2 or along line 3?

**4.** Would the shell appear to be thicker and brighter along line 2 or along line 3?

**5.** Why does the Ring Nebula appears to us as a ring?

# Star Systems and Galaxies (pp. 117–120)

## Objectives

After completing the lesson, students will be able to

- describe a star system;
- identify and describe the three types of galaxies.

## 1 Engage/Explore

- **Activating Prior Knowledge** Compare optical illusion tricks to optical illusion of stars in constellations and how they appear to be next to each other, but are actually at very different distances from Earth.
- **Discover** Invite students to model looking at many distant stars that are close together using white paper with many holes set against a dark surface.

## 2 Facilitate

- **Using the Visuals: Figure 14** Have students describe how eclipsing binaries appear from Earth. Encourage students to draw diagrams showing the positions of Algol, its companion star, and Earth during an eclipse of the star.
- **Inquiry Challenge** Challenge students to design models to represent discovering planets near stars. Ask them to describe their observations of the planet models.
- **Integrating Life Science** Help students understand why scientists are looking for radio signals from space.
- **Cultural Diversity** Encourage students to list ten items representative of their own culture they would like to bring with them on a journey to an extraterrestrial civilization to give a glimpse of life on Earth.
- **Building Inquiry Skills: Calculating** Reinforce the large magnitude of one trillion by having students predict how long it would take for a trillion seconds to pass.
- **Try This** Allow students to make a model of our galaxy using pipe cleaners. Ask them if it's possible to get a good view of stars in the spiral arms from Earth.

## 3 Assess

- **Section 4 Review** Check students' understanding of star systems and galaxies by having them answer the section review questions.
- **Performance Assessment** Assess students' understanding of the three types of galaxies by having them draw three types of galaxies, then write captions describing the features of each.

3

Name ______________________ Date __________ Class ______________

SECTION 3-4 SECTION SUMMARY

# Star Systems and Galaxies

**Guide for Reading**

- What is a star system?
- What are the three types of galaxies?

Our solar system has only one star, the sun. **But more than half of all stars are members of groups of two or more stars, called star systems.** Star systems with two stars are called double stars or **binary stars.** Those with three stars are called triple stars.

Astronomers can sometimes find a binary star even if only one of the stars in the pair can be seen from Earth. The darker star in the pair may pass in front of the other star and block the light from the other star. A system in which one star blocks the light from another is called an **eclipsing binary.** Often astronomers can tell that there is a second star in a system by observing the effects of its gravity.

Astronomers have found planets revolving around stars by using a method like the one they use to find binary stars. All the planets found beyond our solar system so far are very large. A small planet would be difficult to find because it would have little gravitational pull on the star it revolved around. Astronomers are trying new ways to use telescopes to see planets directly. To see a planet directly, they must shield their view of the planet from the glare of the star.

Some scientists think that life may exist on planets revolving around other stars. A few astronomers are using radio telescopes to search for signals that could not come from natural sources. Such a signal might be evidence of extraterrestrial civilizations.

The Milky Way is the galaxy in which our solar system is located. Like other galaxies, it has single stars, double stars, star systems, and lots of gas and dust between the stars. The Milky Way Galaxy looks milky or hazy because stars are too close together for your eyes to see them individually. The dark blotches in the Milky Way are clouds of dust that block light from behind them.

**Astronomers have classified galaxies into three main categories: spiral galaxies, elliptical galaxies, and irregular galaxies.** A **spiral galaxy** is a galaxy with arms that spiral outward, like a pinwheel. **Elliptical galaxies** look like flattened balls. They contain billions of stars but have little gas and dust between stars. Some galaxies do not have regular shapes, so they are known as **irregular galaxies.**

Name ______________________ Date __________ Class __________

SECTION 3-5 REVIEW AND REINFORCE

# History of the Universe

## ◆ Understanding Main Ideas

*Write an answer for each of the following questions in the spaces provided.*

1. In which direction are nearly all galaxies moving relative to Earth?

______________________________________________

2. What is the general relationship between a galaxy's distance from Earth and its speed?

______________________________________________

*Place the following events in the order in which they would occurred. Place the order number (1 through 7) in the blank to the left of each event.*

3

________ **3.** nebula shrinks to form a spinning disk

________ **4.** gravity pulls gases to center of the disk

________ **5.** big bang

________ **6.** the planets form

________ **7.** gas and dust accumulate as a nebula

________ **8.** matter separates into galaxies

________ **9.** gas at the center of the disk becomes hot and dense enough for nuclear fusion to begin

## ◆ Building Vocabulary

*Write an answer for the following question in the space provided.*

**10.** What is the big bang theory?

______________________________________________

______________________________________________

______________________________________________

Name ______________________ Date __________ Class ______________

**SECTION 3-5** **ENRICH**

# How Far and How Fast?

You know that electromagnetic radiation is energy in the form of waves. Planets, stars, and galaxies emit electromagnetic radiation. As one of these objects moves through space toward an observer, the waves of electromagnetic radiation between the object and the observer move closer together. That is, the distance between the crests of the waves decreases, causing the wavelength to shorten. As that same object moves away from an observer, the waves between the object and the observer move farther apart from each other. The distance between the crests of the waves increases, and the wavelength lengthens.

Figure 1

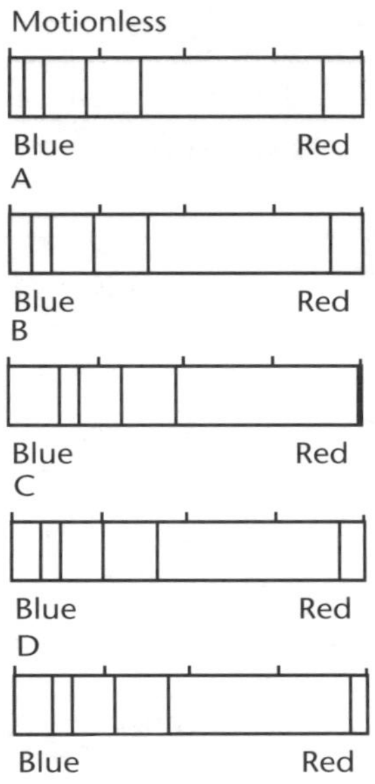


3

When a star is moving toward Earth very fast, the wavelength of its light shortens, causing it to appear a little more blue than usual. If a star is moving away from Earth very fast, the wavelength of its light lengthens, causing it to appear a little more red. By measuring this *blue shift* or *red shift* of light, astronomers can determine whether the star is moving toward or away from Earth, and how fast. To do this, astronomers use the shift of spectral lines produced by chemicals in the star's atmosphere. Astronomers can also use the spectrum of a galaxy to tell how fast it is moving toward or away from Earth.

Using the principle of red shift, Edwin Hubble studied the relationship between a galaxy's distance and the speed at which it moves away from Earth. He found that the farther away a galaxy is, the faster it moves away from Earth. This relationship is called the *Hubble Law.*

*Based on the spectrums in Figure 1, rank the four galaxies in order of the speed with which they are moving away from Earth, from slowest (1) to fastest (4).*

| | **Rank (1 = slowest, 4 = fastest)** |
|---|---|
| Galaxy A | |
| Galaxy B | |
| Galaxy C | |
| Galaxy D | |

**1.** Based on the Hubble Law, which of the galaxies is farthest from Earth?

**2.** Are any of the galaxies moving toward Earth? Explain.

# Make Your Own Telescope

In this lab, you will learn how to construct and use a simple refracting telescope. You can then try out your telescope.

◆ **Problem** How can you build a telescope?

◆ **Skills Focus** making models, observing, drawing conclusions

◆ **Materials**

2 paper towel tubes of slightly different diameters
plastic objective lens (43-mm diameter, 400-mm focal length)
plastic eyepiece lens (17.5-mm diameter, 25-mm focal length)
foam holder for eyepiece
transparent tape
meter stick

◆ **Procedure**

1. Fit one of the paper towel tubes inside the other. Make sure you can move the tubes but that they will not slide on their own.

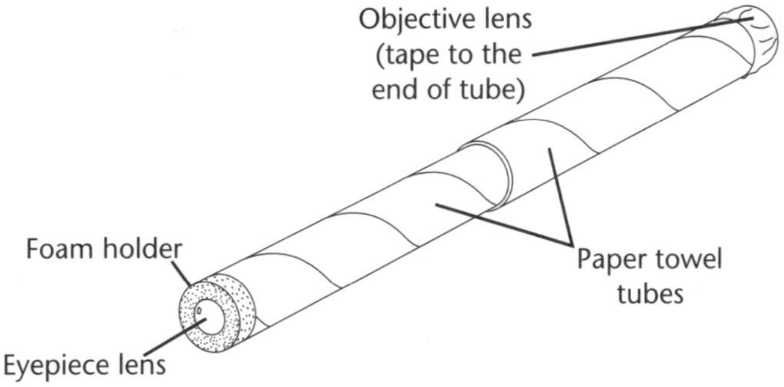


2. Place the large objective lens flat against the end of the outer tube. Tape the lens in place. Try to block as little of the lens with the tape as possible.
3. Make sure the center of the foam holder is free of foam. Insert the small eyepiece lens into the opening in the holder.
4. Place the foam eyepiece lens holder into the inner tube at the opposite end of the telescope to the objective lens.

## REAL-WORLD LAB *(continued)*

5. Tape a meter stick to the wall. Look through the eyepiece at the meter stick from 5 m away. Slide the tubes in and out to focus your telescope so that you can clearly read the numbers on the meter stick.
6. Use your telescope to look at other objects at different distances, both in your classroom and through the window. **CAUTION:** *Do not look at the sun. You will damage your eyes.*

### ◆ Analyze and Conclude

*Write your answers on the back of this sheet or on a separate sheet of paper.*

1. Why do you need two tubes?
2. If you focus on a nearby object and then focus on something farther away, do you have to move the tubes together or apart?
3. How does this telescope compare to the telescopes astronomers use?
4. **Apply** How could you improve on the design of your telescope? What effects would different lenses or tubes have?

3

### ◆ More to Explore

With an adult, go outside in the evening a few days after the first-quarter phase and observe the moon. Point the telescope at the moon and draw a circle with all the features you see. Label the maria (lowlands) and highlands.

Name ______________________ Date __________ Class ______________

CHAPTER 3 SKILLS LAB

*Measuring*

# How Far Is That Star?

When astronomers measure parallax, they record the positions of stars on film in cameras attached to telescopes. In this lab, you will set up a model of a telescope and use it to estimate distances.

## ◆ Problem

How can parallax be used to determine distances?

## ◆ Materials

| | | |
|---|---|---|
| masking tape | paper clips | pen |
| black and red pencils | metric ruler | paper |
| extension cord | meter stick | calculator |

lamp without a shade, with 100-watt light bulb
copier paper box (without the lid)
flat rectangular table, about 1 m wide

## ◆ Procedure

*Review the safety guidelines in Appendix A.*

### Part 1 Telescope Model

1. Place the lamp on a table in the middle of the classroom.
2. Carefully use the tip of the pen to make a small hole in the middle of one end of the box. The box represents a telescope.
3. At the front of the classroom, place the box on a flat table so the hole points toward the light. Line the left side of the box up with the left edge of the table.
4. Put a small piece of tape on the table below the hole. Use the pen to make a mark on the tape directly below the hole. The mark represents the position of the telescope when Earth is on one side of its orbit.

### Part 2 Star 1

5. Label a sheet of paper Star 1 and place it inside the box as shown in the drawing. Hold the paper in place with two paper clips. The paper represents the film in a telescope.
6. Darken the room. Turn on the light to represent the star.
7. With the red pencil, mark the paper where you see a dot of light. Label this dot A. Dot A represents the image of the star on the film.



3

Name ______________________ Date __________ Class ______________

## SKILLS LAB *(continued)*

8. Move the box so the right edge of the box lines up with the right edge of the table. Repeat Step 4. The mark on the tape represents the position of the telescope six months later, when Earth is on the other side of its orbit.
9. Repeat Step 7, and use a black pencil to mark the second dot B. Dot B represents the image of the star as seen 6 months later from the other side of Earth's orbit.
10. Remove the paper. Use the data table provided to record the data that you will collect in the following steps.
11. Measure and record the distance in millimeters between dots A and B. This distance represents the parallax shift of Star 1.
12. Measure and record the distance from the hole (lens) to the paper (film) at the back of the box in millimeters. This distance represents the focal length of your telescope.
13. Measure and record the distance in millimeters between the marks on the two pieces of masking tape. This distance represents the diameter of Earth's orbit.



### Part 3 Stars 2 and 3

14. Move the lamp away from the table—about half the distance to the back of the room. The bulb now represents Star 2. Predict what you think will happen to the light images on your paper.
15. Repeat Steps 6–11 with a new sheet of paper to find the parallax shift for Star 2.
16. Move the lamp to the back of the classroom. The bulb now represents Star 3. Repeat Steps 6–11 with a new sheet of paper to find the parallax shift for Star 3.

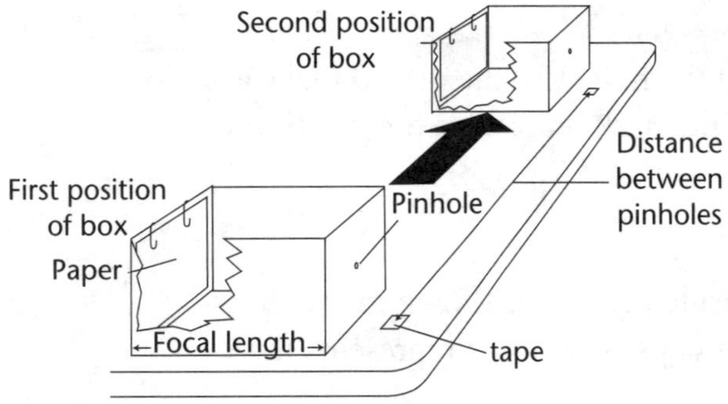


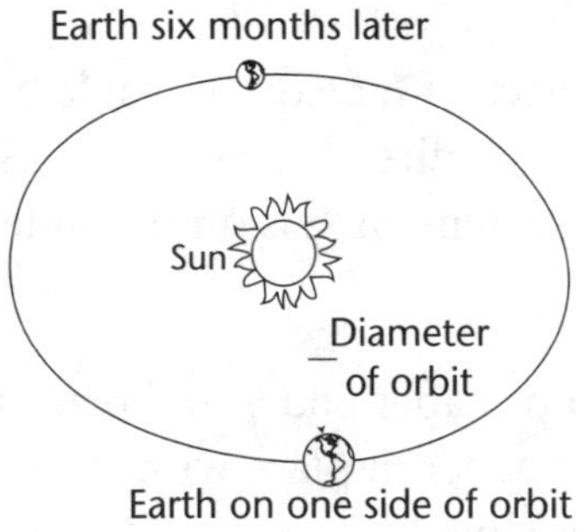

**SKILLS LAB** *(continued)*

## ◆ Data Table

| Star | Parallax Shift (mm) | Focal Length (mm) | Diameter of Orbit (mm) | Calculated Distance to Star (mm) | Calculated Distance to Star (m) | Actual Distance to Star (m) |
|---|---|---|---|---|---|---|
| 1 | | | | | | |
| 2 | | | | | | |
| 3 | | | | | | |

## ◆ Analyze and Conclude

*Write your answers on the back of this sheet or on a separate sheet of paper.*

3

1. What happened to the dot of light for each star when you moved your model telescope from one side of Earth's orbit to the other?
2. What caused the apparent change in position of the dots of light for each star? Explain.
3. Use the following formula to calculate the distance from the telescope to Star 1.

$$\text{Distance} = \frac{\text{Diameter} \times \text{Focal length}}{\text{Parallax shift}}$$

4. Divide your result from Question 3 by 1,000 to get the distance to the light bulb in meters.
5. Repeat Questions 3 and 4 for Stars 2 and 3.
6. Was your prediction in Step 14 correct? Why or why not?
7. Is the parallax shift greater or smaller the farther away the star is? Relate each star's parallax shift to its distance from Earth.
8. **Think About It** Use a meter stick to measure the actual distance from the box to the bulb. How did your calculation for Star 3 compare with the actual distance? What could you do to improve your results?

## ◆ Design an Experiment

What would happen if you kept moving the lamp away from the box? Is there a distance at which you can no longer find the distance to the star? Design an experiment to find out.

Name ______________________ Date __________ Class ______________



# Ain't I A Woman?

*In 1851, Sojourner Truth was attending a Woman's Convention in Akron, Ohio. She contributed an impromptu speech to the proceedings in response to sexist remarks she was hearing. Read these excerpts from the speech and answer the following questions on the lines below.*

That man over there says that women need to be helped into carriages, and lifted over ditches, and to have the best place everywhere. Nobody ever helps me into carriages, or over mud-puddles, or gives me any best place! And ain't I a woman? Look at me! Look at my arm! I have ploughed and planted, and gathered into barns, and no man could head me! And ain't I a woman? I could work as much and eat as much as a man—when I could get it—and bear the lash as well! And ain't I a woman? I have borne thirteen children, and seen most all sold off to slavery, and when I cried out with my mother's grief, none but Jesus heard me! And ain't I a woman?

**1.** In the first paragraph, what generalizations has the man given for why women can't have as many rights as men?

______________________________________________

______________________________________________

**2.** How does Sojourner Truth respond to these reasons?

______________________________________________

______________________________________________

Name ______________________ Date __________ Class ______________

INTERDISCIPLINARY EXPLORATION LANGUAGE ARTS

# Words from Mars Fiction

*In 1898, H. G. Wells wrote a science-fiction story called "War of the Worlds." In his story, creatures from Mars come to Earth and try to capture the planet. Read the opening paragraph of this famous story. Then read the word definitions. Under each definition, write the underlined word from the excerpt that matches the definition.*

No one would have believed in the last years of the nineteenth century that this world was being watched keenly and closely by intelligence greater than man's and yet as mortal as his own; that as men busied themselves about their various concerns they were scrutinized and studied, perhaps almost as narrowly as a man with a microscope might scrutinize the transient creatures that swarm and multiply in a drop of water. With infinite complacency men went to and fro over this globe about their little affairs, serene in their assurance of their empire over matter. It is possible that the infusoria under the microscope do the same. No one gave a thought to the older worlds of space as sources of human danger, or thought of them only to dismiss the idea of life upon them as impossible or improbable. It is curious to recall some of the mental habits of those departed days. At most terrestrial men fancied there might be other men upon Mars, perhaps inferior to themselves and ready to welcome a missionary enterprise. Yet across the gulf of space, minds that are to our minds as ours are to those of the beasts that perish, intellects vast and cool and unsympathetic, regarded this earth with envious eyes, and slowly and surely drew their plans against us. And early in the twentieth century came the great disillusionment.

1. to examine closely and minutely

______________________________

2. to have formed an idea about something on the basis of inadequate evidence

______________________________

3. smug, content, and indifferent

______________________________

4. minute organisms

______________________________

5. calm and unruffled

______________________________

6. passing especially quickly in and out of existence

______________________________

# Move to Mars!

*Work with a partner or in small groups.*

*Imagine that a space agency develops a way to build a space colony on Mars. Scientists from Earth will spend a year traveling to Mars, will live in the Mars colony for five years, and will spend a year returning to Earth. You have been hired by the space agency to recruit scientists to live in the colony.*

**1.** First decide whether you think many people will want to join this program or whether you will need to recruit them. Your decision will determine whether your ad will be trying to attract more people or to screen out unsuitable people. Briefly explain how you made your decision.

______________________________________________

______________________________________________

**2.** With your partner or group, brainstorm the kinds of personal and social skills that scientists will need in order to be successful. Remember that they will be away from Earth for 7 years.

______________________________________________

______________________________________________

**3.** Now brainstorm the kinds of educational backgrounds that the scientists should have.

______________________________________________

______________________________________________

**4.** What kinds of professional experience should your scientists have? How many years?

______________________________________________

______________________________________________

**5.** Skim some of the job advertisements in your local newspaper or on the Internet. What other requirements do the advertisements list that you want to include in your ad?

______________________________________________

______________________________________________

**6.** On a separate sheet of paper, write your job advertisement, using the notes you have made.

INTERDISCIPLINARY EXPLORATION SCIENCE

# Gravity on Mars

*Sir Isaac Newton first wrote a formula that describes gravity. Newton's law established that there is a force of gravity between any two masses. We do not feel the force of gravity from objects around us because they are so tiny. Only a mass the size of a moon, a planet, or a star is large enough to have a force of gravity that we can detect. Newton's law stated that the force of gravity depended only on the masses of the two objects and the distance between them. The further the distance, the weaker the force of gravity. Using simplified versions of Newton's law, you will compare the forces of gravity on Earth and Mars. Answer the following questions on the lines below.*

**1.** A simplified version of Newton's law for Earth is $F = 9.8m$. In this equation, $F$ is the force of gravity felt and $m$ is the mass of the person on the surface of Earth. Calculate what force a person with a mass of 80 kilograms would feel. The units of force are newtons.

______________________________________________

**2.** If the person is 10,000 kilometers above the surface of Earth, the formula changes to $F = 1.5m$. Calculate what force a person with a mass of 80 kilograms would feel 10,000 kilometers above the surface of Earth.

______________________________________________

**3.** A simplified version of Newton's law for the surface of Mars is $F = 4.0m$. Calculate what force a person with a mass of 80 kilograms would feel on the surface of Mars.

______________________________________________

**4.** If the person is 10,000 kilometers above the surface of Mars, the formula changes to $F = 0.25m$. Calculate what force a person with a mass of 80 kilograms would feel 10,000 kilometers above the surface of Mars?

______________________________________________

**5.** Calculate the forces that a person with a mass of 50 kilograms would feel on the surface of Earth and on the surface of Mars.

______________________________________________

______________________________________________

**6.** How does the force of gravity on Mars compare to that on Earth?

______________________________________________

**7.** How high can you jump on Earth? How high do you think you could jump on Mars?

______________________________________________

______________________________________________

Name ______________________ Date __________ Class ______________



# The Polar Caps of Mars

*Read the following paragraph on the polar caps of Mars. Then answer the questions on the lines below.*

As early as 1898, scientists were disagreeing on whether the polar caps of Mars were made of frozen water ice or frozen carbon dioxide. The astronomer William Herschel believed the caps were water ice, like those on Earth. In 1948 and again in 1966, scientists analyzed the spectrums reflected from the caps and concluded that the caps contained water ice rather than carbon dioxide.

In 1966, numerical models of carbon dioxide polar caps and polar caps of water ice were calculated. The behavior of the polar caps on Mars seemed to match the models for carbon dioxide polar caps. The models showed how the caps grew and declined. Data from the *Mariner 6* and *7* spacecraft, a decade later, supported the numerical models. The *Mariner* craft performed spectral analysis on portions of the spectrum that Earth-based astronomers could not see because Earth's atmosphere absorbed these wavelengths.

More recent spacecraft observations show that the polar caps are different from each other. The northern polar cap contains more water ice. The southern polar cap contains mostly carbon dioxide ice.

**1.** What did early scientists assume the polar caps had in common?

______________________________________________

**2.** From where in the solar system did scientists conduct their spectral analyses in 1948? How do you know?

______________________________________________

**3.** What information did the scientists miss in 1948 and 1966,that caused them to believe that the caps contained water ice?

______________________________________________

**4.** What can you infer about the results of numerical models in 1966 for polar caps made of water ice?

______________________________________________

**5.** The southern polar cap disappears completely in the southern hemisphere summer. The northern polar cap shrinks, but does not disappear. How can this be explained by a combination of water ice and carbon dioxide ice.

______________________________________________

______________________________________________

**6.** Do you think a colony on Mars should be nearer the north polar cap or the south polar cap? Explain.

______________________________________________

Name ______________________ Date __________ Class ______________

INTERDISCIPLINARY EXPLORATION SCIENCE

# History of Mars Exploration

*Using an encyclopedia or other reference books, make a timeline showing the history of our exploration of Mars. A few key dates and events are provided.*

## ◆ Materials

encyclopedia or other reference books

| Date | Event |
|---|---|
| | first launch of a liquid fuel rocket |
| | |
| | National Aeronautics and Space Administration (NASA) created |
| | |
| 1961 | |
| | |
| | first successful fly-by of Mars |
| | |
| 1969 | |
| | |
| 1976 | |
| | |
| | |
| | |
| | |
| | |
| | |
| | |

Name ______________________ Date __________ Class __________

INTERDISCIPLINARY EXPLORATION SOCIAL STUDIES

# Earth from Space

*Since astronauts and cosmonauts had their first glimpse of the Earth from space, they have taken thousands of photographs and made notes of their impressions. Research what kinds of natural features and man-made features can be seen from orbit.*

**1.** What kinds of weather conditions can astronauts see from orbit? Can astronauts tell where it is raining? Can they see lightning? If so, what does it look like?

______________________________

______________________________

**2.** Can astronauts tell where the worst air pollution is occurring? What does smog look like from space?

______________________________

______________________________

**3.** How large does a forest fire have to be for it to be visible from space? Have astronauts ever seen the glow from fires or do they see only smoke?

______________________________

______________________________

**4.** What does an erupting volcano look like from space? Have astronauts been able to photograph the active volcano in Hawaii?

______________________________

______________________________

**5.** What city features are discernible from space? Can astronauts see airports? Can they see major highway intersections? What about famous landmarks? Can they see the Statue of Liberty? the Eiffel Tower? the Golden Gate Bridge? How do these features appear from space?

______________________________

______________________________

**6.** What part of Earth is brightest at night? Where are the most lights concentrated? Do some cities seem to have brighter lights than others? What regions of Earth have the fewest lights?

______________________________

______________________________

Name ______________________ Date __________ Class ______________

**INTERDISCIPLINARY EXPLORATION SOCIAL STUDIES**

# The Search for Life on Mars

*In 1877, an Italian astronomer named Giovanni Schiaparelli announced that he could see channels on Mars. His announcement began decades of interest and belief by some that there is life on Mars. In the space below, make a timeline from 1877 showing some key dates in the history of the search for life on Mars. Some key dates, figures, and spacecraft are provided to give you clues.*

*Briefly describe what future missions to Mars are planned.*

| Date | Event |
|---|---|
| 1877 | Schiaparelli announces that Mars has channels. The Italian word for channel "canali" was misinterpreted by some to mean "canals." |
| 1894 | Percival Lowell |
| | Antoniadi |
| 1964 | *Mariner 4* spacecraft |
| 1969 | *Mariners 6* and *7* |
| 1971 | *Mariner 9* |
| 1976 | *Vikings* |
| | |
| 1992 | *Mars Observer* |
| | |
| 1996 | *Mars Global Surveyor* |
| | |
| 1997 | *Pathfinder* |
| | |
| 2001? | |
| 2005? | |
| 2009? | |

Name ______________________ Date __________ Class ______________



# Map of Mars

*In the space below, sketch a map showing a region of Mars and some of its geologic features. Label the features with their names. North should be toward the top of the page and the space should represent an area around 500–1000 km wide. Include a crater and other interesting features.*

Name ______________________ Date __________ Class ______________

INTERDISCIPLINARY EXPLORATION MATHEMATICS

# A Year on Mars

*In the space below, design a calendar for the Martian year. Decide how many months the calendar will have, how many sols (Martian days) are in each month, and how many sols are in a week. Name the months. Since the Martian sol is just slightly longer than an Earth day, you can choose either to keep the familiar names of days or think of new Martian names. Finally, figure out when in the calendar the seasons will occur. Remember that the seasons for the northern and southern hemispheres differ.*

Name ________________________ Date __________ Class __________

INTERDISCIPLINARY EXPLORATION MATHEMATICS

# Model the Mars Neighborhood

*You will design a three-dimensional scale model of the inner solar system. Follow the steps to make your model. Your model can be small enough to fit in a shoe box or large enough to cover a football field. You do not need to use the same scale for distance and size.*

**1.** Find out the names of the three planets closest to the sun and the names of each moon. For each planet, find out its diameter and its distance from the sun. For each moon, find out its diameter and its distance from the planet it orbits. Record your findings on the lines below.

______________________________________________

______________________________________________

**2.** Now you will decide the size of your model. Look at the largest and smallest distances. If you use a scale of 1 centimeter = 1,000 kilometers, what will be approximately the smallest and largest distances in your model?

______________________________________________

______________________________________________

**3.** If you want your model to be smaller, change the scale so that 1 centimeter represents more kilometers. If you want your model to be larger, change the scale so that 1 centimeter represents fewer kilometers. Once you have a scale you like, calculate what approximate distance in your model will represent each of the distances between the planetary bodies.

______________________________________________

______________________________________________

**4.** Now you will decide the size of your objects in your model that will represent the planets and moons. Look at the largest and smallest diameters. If you use a scale of 1 centimeter = 100 kilometers, what will be approximately the smallest and largest diameters in your model?

______________________________________________

______________________________________________

**5.** If you want your model to be smaller or larger, change it as you did in step 3. Once you have a scale you like, calculate what approximate diameters in your model will represent each of the diameters of the planets and moons.

______________________________________________

______________________________________________

**6.** Now that you have made all your calculations, you are ready to build your model.

Name ______________________ Date __________ Class ______________

INTERDISCIPLINARY EXPLORATION MATHEMATICS

# Martian Seasons in Days

*The table on page 132 of the text shows the number of sols that each season lasts in the northern and southern hemispheres of Mars. Knowing that one Martian year is equivalent to 687 days, make a table showing the length of the seasons in Earth days for both hemispheres. Write the number of days, rounded to the nearest day. Then write how many months and days this is, approximating a month as 30 days. Use the space below for your calculations.*

**Martian Seasons in Days**

| | Northern Hemisphere (days) | Northern Hemisphere (months and days) | Southern Hemisphere (days) | Southern Hemisphere (months and days) |
|---|---|---|---|---|
| Winter | | | | |
| Spring | | | | |
| Summer | | | | |
| Fall | | | | |

1

# CHAPTER 1 The Tides

Students are asked the question, "How can you construct a model to illustrate how the position of Earth, the sun, and the moon affect the tides?" To answer this question, students will use the balls to represent Earth, the sun, and the moon, and to use the clay as tides on Earth. Students should draw on their knowledge of chapter concepts to manipulate the model and answer the question.

## ◆ Expected Outcome

Students should use the tennis ball, golf ball, and marble to represent the sun, Earth, and the moon. Modeling clay should be used to represent tides on Earth. An important concept for students to demonstrate is that at all times on Earth, two sides of Earth experience high tides and two sides of Earth experience low tides. When the sun, Earth, and the moon are in a line (in two different ways), tides are most extreme (spring tides). When the moon is at right angles to a line connecting the sun and Earth, tides are least extreme (neap tides).

## ◆ Content Assessed

- This activity assesses students' understanding of how the movements of the sun, Earth, and moon affect the tides and phases of the moon.

## ◆ Skills Assessed

making models, applying concepts, communicating

## ◆ Content Assessed

- This activity assesses students' understanding of the movements of Earth, the sun, and the moon and the results of those movements.

## ◆ Materials

- Tennis ball, golf balls, marbles, and modeling clay

## ◆ Advance Preparation

- Acquire enough tennis balls, golf balls, and marbles so that each student can have one of each.
- Prepare about 10 $cm^3$ of modeling clay for each student.

## ◆ Time

One class period

## ◆ Monitoring the Task

- If students have trouble deciding how to begin, suggest that the golf ball, marble, and tennis ball be used to represent Earth, the moon, and the sun, respectively.
- Students will be able to demonstrate their knowledge of chapter concepts best if each student can individually show and explain to you how the relative positions of the parts of the model answer the questions.

## ◆ Extension

- As an extension for this performance assessment, you could model the movement of tides through one day—the Earth seems to rotate beneath the tides, while the high tides stay on the side of Earth closest to and farthest from the moon. You could illustrate this difficult concept by building a sleeve of posterboard around the golf ball, and to layer the clay on the posterboard. The posterboard tube can move independently of the golf ball beneath, as the height of tides does on Earth.
- Students could also use their models to demonstrate their understanding of the mechanisms behind seasons and eclipses.

CHAPTER 1

# The Tides

In assessing students' performance, use the following rubric.

| | 4 | 3 | 2 | 1 |
|---|---|---|---|---|
| ***Constructing a Model*** | Student's model correctly shows the relative positions of Earth, the sun, and the moon, and that two simultaneous high and low tides change position independently of the rotation of Earth. | Student's model correctly shows the relative positions of Earth, the sun, and the moon, and that there are two simultaneous high and low tides. | Student's model has only a few errors in the relative positions of Earth, the sun, and the moon, but shows that there are two simultaneous high and low tides. | Student's model has many errors in the relative positions of Earth, the sun, and the moon, and incorrectly models high and low tides. |
| ***Concept Understanding*** | Student demonstrates thorough understanding of how the positions of Earth, the sun, and the moon affect tides and moon phases. | Student demonstrates adequate understanding of how the positions of Earth, the sun, and the moon affect tides and moon phases. Student has some minor misunderstandings of concepts. | Student demonstrates some understanding of how the positions of Earth, the sun, and the moon affect tides and moon phases, but several important concepts are not understood. | Student has serious misunderstandings of how the positions of Earth, the sun, and the moon affect tides and moon phases. |

1

# The Tides

◆ **Problem** How can you construct a model to illustrate how the position of Earth, the sun, and the moon affect the tides?

## ◆ Suggested Materials

tennis ball
golf ball
marble
modeling clay

## ◆ Devise a Plan

1. Study the materials and devise a plan to model how the positions of Earth, the sun, and the moon affect the Earth's tides.
2. Write down your plan. You will probably want to include sketches.

## ◆ Analyze and Conclude

*After performing the activity, answer the following questions on a separate sheet of paper.*

1. Use your model to show the positions of Earth, the sun, and the moon when Earth has spring tides. Also show the position of the high and low tides on Earth at those positions.
2. In what phase is the moon in each of the positions in question 1?
3. Use your model to show the positions of Earth, the sun, and the moon when Earth has neap tides. Also show the position of the high and low tides on Earth at those positions.
4. In what phase is the moon in each of the positions in question 3?

1

# The Earth, Moon, Sun

## ◆ Multiple Choice

*Write the letter of the correct answer on the line at the left.*

______ **1.** Earth's spinning on its axis is called its
**a.** revolution. **b.** orbit.
**c.** rotation. **d.** cycle.

______ **2.** Which of the following events occurs once every 24 hours?
**a.** Earth revolves around the sun. **b.** The moon rotates on its axis.
**c.** The moon revolves around Earth. **d.** Earth rotates on its axis.

______ **3.** The measurement of distance from the equator is
**a.** circumference. **b.** latitude.
**c.** longitude. **d.** degrees.

______ **4.** The moon's orbit around Earth has the shape of a(n)
**a.** circle. **b.** oval.
**c.** sphere. **d.** crescent.

______ **5.** Spring tides occur
**a.** when the sun and moon are at right angles to Earth.
**b.** when day and night are equal.
**c.** when the sun, moon, and Earth line up.
**d.** only when the sun and moon are the same distance from Earth.

______ **6.** A rocket is pushed forward as a reaction to
**a.** the expelling of hot gases through a nozzle at the rear.
**b.** the expelling of rocket fuel through a nozzle at the rear.
**c.** the heat of burning rocket fuel.
**d.** the gravity of Earth.

______ **7.** The pull of gravity on the surface of the moon is
**a.** six times that of Earth. **b.** one quarter that of Earth.
**c.** one eightieth that of Earth. **d.** one sixth that of Earth.

______ **8.** Maria are
**a.** seas on the moon.
**b.** regions with many craters.
**c.** regions once flooded by molten material.
**d.** lunar highlands.

## CHAPTER 1 TEST *(continued)*

______ **9.** The darkest part of the moon's shadow is the
**a.** penumbra. **b.** umbra.
**c.** solar eclipse. **d.** new moon.

______ **10.** Humans first landed on the moon in
**a.** 1861. **b.** 1961.
**c.** 1969. **d.** 1971.



## ◆ Completion

*Fill in the line to complete each statement.*

**11.** The two days on which the sun is overhead at either 23.5 degrees north or south are called ______________.

**12.** If you stand in the ______________, you will see only a partial eclipse.

**13.** Satellites with ______________ orbits appear to hover over one place on Earth's surface.

**14.** An eclipse of the moon appears only when the moon is in the ______________ phase.

**15.** Each of the two days of the year when neither hemisphere is tilted toward the sun is known as a(n) ______________.

## ◆ True or False

*Determine whether each statement is true or false. If it is true, write true. If it is false, change the underlined word or words to make the statement true.*

______________ **16.** In one year, Earth completes 365 <u>revolutions</u>.

______________ **17.** The day in March on which the sun is overhead at the equator is called the <u>autumnal solstice</u>.

______________ **18.** A <u>solar eclipse</u> occurs during a full moon when Earth is directly between the moon and the sun.

______________ **19.** When the moon is in Earth's <u>umbra</u>, you see a total lunar eclipse.

______________ **20.** The first person to orbit Earth was <u>Yuri Gagarin</u>.

CHAPTER 1 TEST *(continued)*

## ◆ Using Science Skills

*Use the diagram below to answer questions 21 and 22. Answer each question on the lines provided or on a separate sheet of paper.*

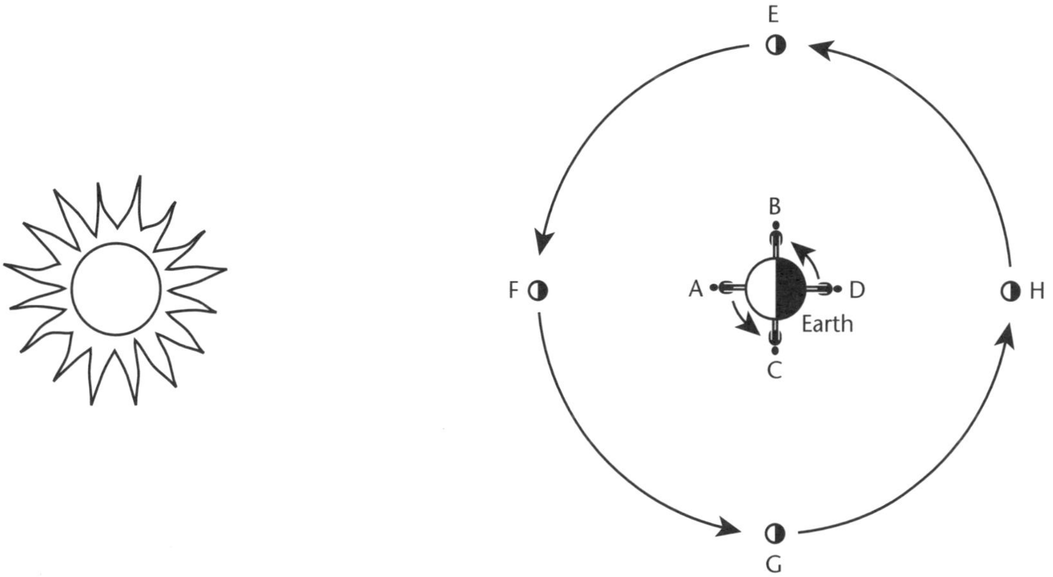


**21. Inferring** Where in the sky would the sun appear to an observer at each of the four positions on Earth shown in the diagram? (Examples: overhead, on the horizon, not visible.)

**A.** ____________________

**B.** ____________________

**C.** ____________________

**D.** ____________________

**22. Drawing Conclusions** What is the phase of the moon when it is in each of the positions shown?

**E.** ____________________

**F.** ____________________

**G.** ____________________

**H.** ____________________

## ◆ Essay

*Write an answer for each of the following questions on a separate sheet of paper.*

**23.** Describe how the tides change at a point on the seashore over the course of a day.

**24.** Explain why Earth experiences seasons.

**25.** Describe the events that most likely created the moon.

1

## ◆ Using Science Skills

*Use the diagram below to answer the following questions. Answer each in a brief paragraph on a separate sheet of paper.*

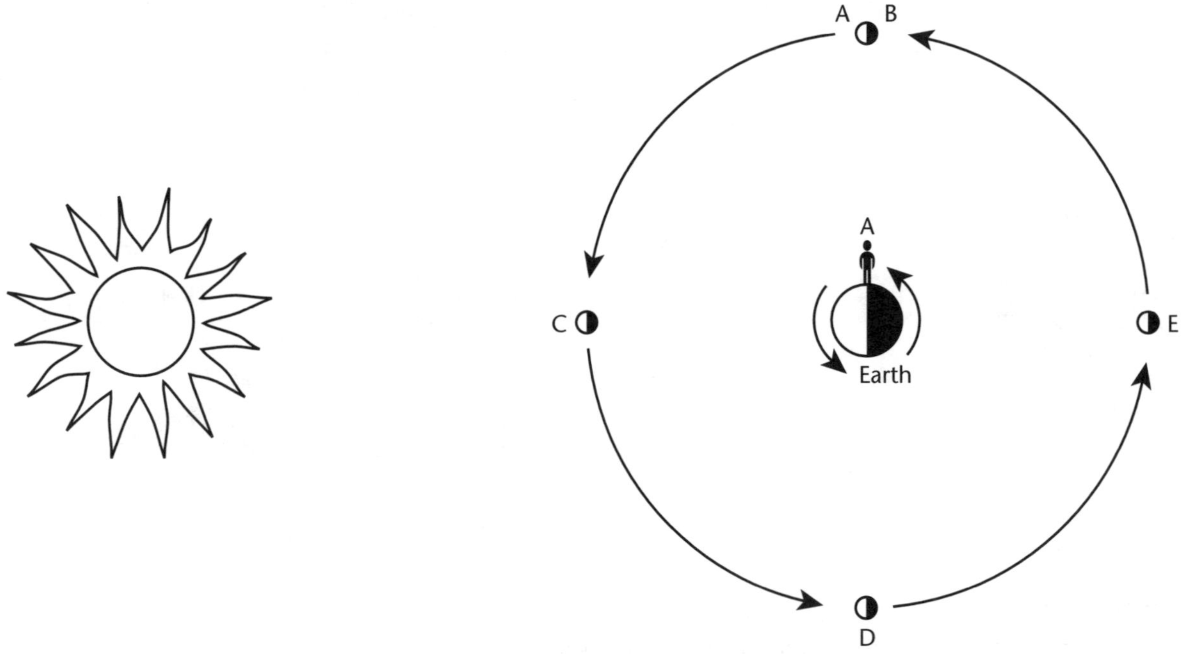


**26. Interpreting Diagrams** In which two positions of the moon would Earth experience neap tides? In which would it experience spring tides? Explain.

**27. Applying Concepts** What phases of the moon would be visible during the highest spring tides? At that time, would the person at point A be witnessing high or low tide?

## ◆ Essay

*Write an answer for each of the following questions on a separate sheet of paper.*

**28.** Why can't astonomers use telescopes to study the entire surface of the moon from Earth? What other methods could they use?

**29.** Explain why we observe solar and lunar eclipses. Which do you think would be more common? Why?

**30.** Describe how the Apollo astronauts studied the moon and some of the things scientists have learned from the Apollo missions.

# CHAPTER 2 Solar System Trading Cards

Students demonstrate their understanding of the solar system by making "trading cards" of the sun and other objects in the solar system.

## ◆ Expected Outcome

Students will select six objects in the solar system. They should pick the sun, an inner planet, an outer planet, and either comets, asteroids, or meteoroids. To these four, students should add two more objects of their choice from any of the groups.

## ◆ Content Assessed

This Performance Assessment assesses students' understanding of the objects in the solar system.

## ◆ Skills Assessed

communicating, classifying

## ◆ Materials

- Each student will need at least six 4 × 6 index cards.

## ◆ Advance Preparation

- You and your students should gather the various materials required to make the trading cards.

## ◆ Time

Allow one class period for this Performance Assessment.

## ◆ Monitoring the Task

- Examine students' plans before they begin.
- At the conclusion of the performance assessment you could assemble a pack of cards that describes the main objects mentioned in the textbook.

CHAPTER 2

# Solar System Trading Cards

In assessing students' performance, use the following rubric.

| | 4 | 3 | 2 | 1 |
|---|---|---|---|---|
| ***Communicating*** | Information on the cards is organized and neat and effectively communicates most of the chapter concepts. | Information on the cards is fairly well organized and neat and communicates many chapter concepts. | Information on the cards is somewhat disorganized but communicates some chapter concepts. | Information on the cards is disorganized and messy and communicates few concepts. |
| ***Concept Understanding*** | Cards indicate superior understanding of both the details of individual objects and the chapter's broad ideas. Each card has more than 5 facts about the object. | Cards indicate understanding of both the details of individual objects and the chapter's broad ideas. Each card has 5 facts about the object. | Cards indicate some understanding of both the details of individual objects and the chapter's broad ideas. A few cards have fewer than 5 facts about the object. | Cards indicate minimal understanding of the details of individual objects but not the chapter's broad ideas. Most cards have fewer than 5 facts about the object. |

Name ______________________ Date __________ Class ______________



# Solar System Trading Cards

You have probably seen various kinds of trading cards. Some of these feature sports figures and include information about the player, statistics relating to his or her skill as an athlete, and a picture. Similar cards exist for action figures.

## ◆ Problem

How can you create a set of trading cards for the objects in our solar system, including the sun, the inner planets, the outer planets, and the comets, asteroids, and meteoroids?

2

## ◆ Suggested Materials

4 × 6 index cards
pictures from magazines
pencils and pens
colored pencils
reference books
tempera paint
textbook

## ◆ Devise a Plan

1. Select a total of six objects from the solar system to investigate. One must be the sun and you must select at least one object from the inner planets, at least one from the outer planets, and at least one from comets, asteroids, and meteoroids.
2. Study the materials and think of a way they can be used to create trading cards (one card for each of six objects that you select).
3. Gather information for your cards. A table is a good way to organize the information. Each card needs at least five facts about that object. One goal should be to include enough information that someone could figure out what the object is without knowing its name.
4. Make a design for your cards.
5. Have your plan approved by the teacher.
6. Modify your plan if necessary and complete your cards.



## ◆ Analyze and Conclude

*After creating your set of cards, answer the following questions on the back of this sheet or on a separate sheet of paper.*

1. Which objects did you select? What information did you include on each card?
2. Some information about your objects might be very difficult to include on a card. What is some information you could not include on your cards and why couldn't you include it?

Name ______________________ Date __________ Class ______________

CHAPTER 2 CHAPTER TEST

# The Solar System

## ◆ Multiple Choice

*Write the letter of the correct answer on the line at the left.*

______ **1.** A sun-centered system of planets was developed by
**a.** Galileo. **b.** Ptolemy.
**c.** Copernicus. **d.** Isaac Newton.

2

______ **2.** The flash of light produced when a meteoroid passes through Earth's atmosphere is a(n)
**a.** meteorite. **b.** comet.
**c.** asteroid. **d.** meteor.

______ **3.** The four terrestrial planets are known for their
**a.** hot temperatures. **b.** rocky surfaces.
**c.** deep atmospheres. **d.** large size.

______ **4.** A stream of particles that are produced by the sun's corona is called
**a.** momentum. **b.** chromosphere.
**c.** prominence. **d.** solar wind.

______ **5.** The planet Venus rotates east to west, which means it has
**a.** retrograde rotation. **c.** no sunrise.
**b.** almost no atmosphere. **d.** a tilted axis.

______ **6.** Which of the following is *not* included in scientists "Goldilocks conditions" for supporting life.
**a.** liquid water **c.** ice
**b.** appropriate temperature range **d.** favorable atmosphere

______ **7.** The _______ is between the orbits of Mars and Jupiter.
**a.** solar system **c.** terrestrial planets
**b.** asteroid belt **d.** photosphere

______ **8.** A(n) _______ is a chunk of ice and dust that has a bright head and a long narrow tail.
**a.** asteroid **c.** meteoroid
**b.** planet **d.** comet

## CHAPTER 2 TEST *(continued)*

______ **9.** ______ are loops of gas on the sun that link sunspots together.
- **a.** Prominences
- **b.** Solar flares
- **c.** Magnetic storms
- **d.** Particle streams

______ **10.** ______ are often considered a double planet.
- **a.** Earth and Venus
- **b.** Pluto and Charon
- **c.** Jupiter and Europa
- **d.** Saturn and Titan



## ◆ Completion

*Fill in the line to complete each statement.*

**11.** Kepler used Brahe's observations to show that the orbits of the planets are ______________ .

**12.** ______________ are areas of gas on the sun that are cooler than the gases around them.

**13.** ______________ are meteoroids that entered our atmosphere and hit Earth without completely burning up.

**14.** Scientist believe that Europa, one of Jupiter's moons, may have ______________ beneath its icy surface that could support life.

**15.** Meteoroids usually come from ______________ or asteroids.

## ◆ True or False

*If the statement is true, write true. If it is false, change the underlined word or words to make the statement true.*

____________________ **16.** The tendency of an object to stay in motion or remain at rest is called <u>gravity</u>.

____________________ **17.** A <u>geocentric</u> model of the planets is sun-centered.

____________________ **18.** <u>Solar flares</u> are explosions that occur when gas on the surface of the sun becomes extremely hot.

____________________ **19.** <u>Asteroids</u> are chunks of rock or dust in orbit that are too small and too numerous to be considered planets.

____________________ **20.** The sun's energy comes from nuclear <u>fission</u>.

## ◆ Using Science Skills

*Use the diagram below to answer each of the following*

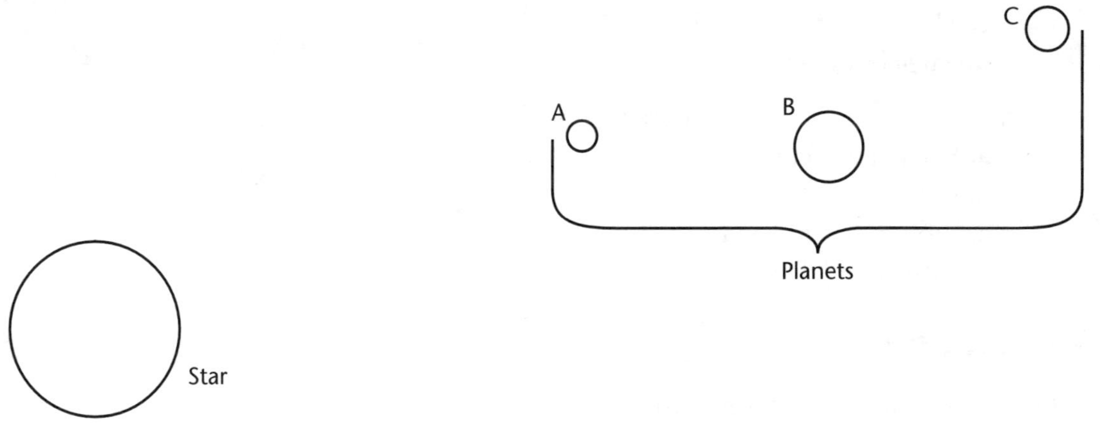


21. **Inferring** Which planet will take the longest time to revolve around the star? Why?

_______________________________________________

_______________________________________________

22. **Interpreting Diagrams** If the planets shown were made of the same material, which one would have the greatest gravitational pull? Why?

_______________________________________________

_______________________________________________

## ◆ Essay

*Write an answer for each of the following.*

23. According to Isaac Newton, why do the planets remain in orbit?

_______________________________________________

_______________________________________________

_______________________________________________

24. Describe how the sun creates energy in its core.

_______________________________________________

_______________________________________________

_______________________________________________

25. List the four gas giants and explain how they are similar to each other and different from the inner planets.

_______________________________________________

_______________________________________________

_______________________________________________

2

## ◆ Using Science Skills

*Use the diagram below to answer each of the following.*

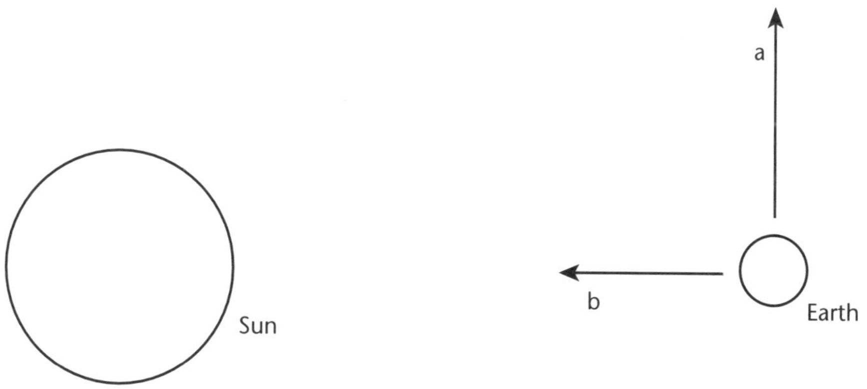


**26. Interpreting Diagrams** Isaac Newton concluded that there were two factors that keep the planets in orbit. Label the arrows on the diagram above to name the factors that keep the planet Earth in orbit.

______________________________________________

______________________________________________

**27. Predicting** On the diagram above, draw a dotted line showing the path you believe Earth would follow if the sun suddenly disappeared.

______________________________________________

______________________________________________

## ◆ Essay

*Write an answer for each of the following.*

**28.** The temperatures on Mercury range from 430°C in the day to –170°C at night. Explain why the temperature varies so much.

______________________________________________

______________________________________________

______________________________________________

**29.** Why do scientists suspect life on Mars might have been possible at one time?

______________________________________________

______________________________________________

______________________________________________

**30.** Explain how the greenhouse effect works on Venus.

______________________________________________

______________________________________________

______________________________________________

CHAPTER 3

# Sizes of Stars



This activity provides an opportunity for students to conceptualize the enormous difference in size between the categories of objects mentioned in this chapter. To do this, they must create a model and apply concepts from the chapter.

## ◆ Expected Outcome

Students' models should use the large beach ball to represent the first object of the pair selected and either the golf ball or the ball bearing to represent the second object. The ratios of diameters are:

| | | | |
|---|---|---|---|
| red giant | : | the sun | 100:1 |
| the sun | : | white dwarf | 200:1 |
| white dwarf | : | neutron star | 500:1 |

The beach ball and the golf ball are a reasonable model for the first two pairs while the beach ball and the ball bearing are an approximate model for the last pair. After creating the model, students are asked to make a diagram showing the four stellar objects with life history information.

## ◆ Content Assessed

This activity assesses students' understanding of the enormous size difference between red giants, medium-sized stars, white dwarfs, and neutron stars. Students' understanding of the characteristics and life cycles of those objects is also assessed.

## ◆ Skills Assessed

building models, inferring, applying concepts

## ◆ Materials

- ◆ large beach ball, 1 m in diameter
- ◆ golf ball or other ball about 2 cm in diameter
- ◆ small ball bearing or small copper shot about 2 mm in diameter

## ◆ Advance Preparation

- ◆ Obtain sufficient beach balls, golf balls, and ball bearings for each student who will be performing the task.

## ◆ Time

20 minutes

## ◆ Safety

Be sure students do not exhibit inappropriate classroom behaviors when using the beach balls.

## ◆ Monitoring the Task

- ◆ Be sure students have decided how to represent objects from each pair.
- ◆ Provide assistance to students who have difficulty translating the information from the chapter into relative sizes. The pertinent sizes are these: red giant—"large enough to fill the solar system as far out as Jupiter," white dwarf—"about the size of Earth," neutron star—"about 20 kilometers in diameter."
- ◆ Students may work in pairs.

# CHAPTER 3 Sizes of Stars

In assessing students' performance, use the following rubric.

| | 4 | 3 | 2 | 1 |
|---|---|---|---|---|
| ***Inferring*** | Inferences about sizes are logical and based on information from the chapter. | Inferences are logical but not based on information. | Inferences are speculative and not based on information. | Inferences do not make sense and are not related to sizes of objects. |
| ***Making Models*** | The models correctly represent relative sizes of objects. The models are correctly labeled. | The models correctly represent the relative sizes of objects. The models are incorrectly labeled or labels are unclear. | The models are correctly labeled but do not correctly represent the relative sizes. | The models are incorrectly labeled and do not correctly represent the relative sizes. |
| ***Applying Concepts*** | Demonstrates thorough knowledge of the relative sizes of the four objects. Diagram demonstrates excellent knowledge of stars' life history and characteristics. | Demonstrates adequate knowledge of the relative sizes of the four objects. Diagram demonstrates basic knowledge of stars' life history and characteristics. | Demonstrates some knowledge of the relative sizes of the four objects. Diagram demonstrates some gaps in knowledge of stars' life history and characteristics. | Demonstrates little knowledge of the relative sizes of the four objects. Diagram demonstrates significant gaps or misconceptions about stars' life history and characteristics. |

Name ______________________ Date __________ Class __________



# Sizes of Stars

## ◆ Problem

How can you model the sizes of stars and other objects?

## ◆ Suggested Materials

large inflatable "beach" ball (about 1 m in diameter)
golf ball or other ball (about 2 cm in diameter)
small ball bearing (2 mm in diameter or smaller)

## ◆ Devise a Plan

1. For each of these pairs of objects, think of a way the materials provided can be used to make a model of the relative sizes of the objects.

| | | |
|---|---|---|
| **A.** | red giant | the sun |
| **B.** | the sun | white dwarf |
| **C.** | white dwarf | neutron star |

2. Make a sketch of the setup you plan to use.
3. Use the materials to construct models of the relative sizes of each pair of objects.

## ◆ Analyze and Conclude

*After following the plan you devised, answer the following questions on a separate sheet of paper.*

1. In each model, what does the small object represent?
2. In each model, what does the large object represent?
3. What does your model tell you about the relative sizes of each pair of objects?
4. For pair A, how could you include a white dwarf in your model? For pair B, how would you include a red giant in your model? For pair C, how would you add the sun to your model?
5. How could you design a single model to include all four objects (red giant, sun, white dwarf, neutron star)?
6. Make a diagram showing how these four objects are related, using arrows to show the order in which the stages occur. Add labels describing what else you know about each stage, such as what factor determines a star's ultimate fate.

Name ______________________ Date __________ Class ____________________

CHAPTER 3 TEST

# Stars, Galaxies, and the Universe

## ◆ Multiple Choice

*Write the letter of the correct answer on the line at the left.*

______ **1.** Patterns of stars in the sky are called
- **a.** galaxies.
- **b.** constellations.
- **c.** star systems.
- **d.** observatories.

______ **2.** Which of the following is *not* part of the electromagnetic spectrum?
- **a.** radio waves
- **b.** sound waves
- **c.** X-rays
- **d.** infrared radiation

______ **3.** A neutron star that produces pulses of radio waves is called a
- **a.** quasar.
- **b.** binary system.
- **c.** black hole.
- **d.** pulsar.

______ **4.** The earliest stage of a star's "life" is called a
- **a.** protostar.
- **b.** giant star.
- **c.** quasar.
- **d.** white dwarf.

______ **5.** The Hertzsprung-Russell diagram graphs stars'
- **a.** brightness and temperature.
- **b.** brightness and absolute magnitude.
- **c.** distance and brightness.
- **d.** diameter and brightness.

______ **6.** How long a star "lives" depends on
- **a.** its color.
- **b.** its mass.
- **c.** whether it is single or binary.
- **d.** its temperature.

______ **7.** Parallax can't be used on a star if the star is too
- **a.** close.
- **b.** bright.
- **c.** far away.
- **d.** large.

______ **8.** A light-year is a unit of
- **a.** time.
- **b.** distance.
- **c.** mass.
- **d.** absolute magnitude.

_____ **9.** The distance between the crest of one wave of electromagnetic radiation and the crest of the next one is its
**a.** frequency.
**b.** wavelength.
**c.** color.
**d.** spectrum.

_____ **10.** When a star runs out of fuel, it *cannot* become a
**a.** protostar.
**b.** neutron star.
**c.** white dwarf.
**d.** black hole.

## ◆ Completion

3

*Fill in the line to complete each statement.*

**11.** A star system in which one star blocks the light from another is called a(n) ________________.

**12.** A device used to detect radio waves from objects in space is called a(n) ________________.

**13.** A(n) ________________ telescope focuses light through a transparent lens.

**14.** A(n) ________________ telescope collects and focuses light using a mirror.

**15.** A giant cloud of gas and dust in the galaxy is called a(n) ________________.

## ◆ True or False

*Determine whether each statement is true or false. If it is true, write true. If it is false, change the underlined word or words to make the statement true.*

____________________ **16.** A star is born when nuclear <u>fusion</u> begins inside a protostar.

____________________ **17.** A quasar is a type of <u>galaxy</u>.

____________________ **18.** A star system containing two stars is called a <u>binary star</u>.

____________________ **19.** A star's <u>absolute magnitude</u> is its brightness as seen from Earth.

____________________ **20.** The hottest stars are <u>blue-white</u> in color.

Name ______________________ Date __________ Class ______________



## ◆ Using Science Skills

*Use the diagram below to answer questions 21–23. Write your answers on the back of this page or on a separate sheet of paper if you need more room.*

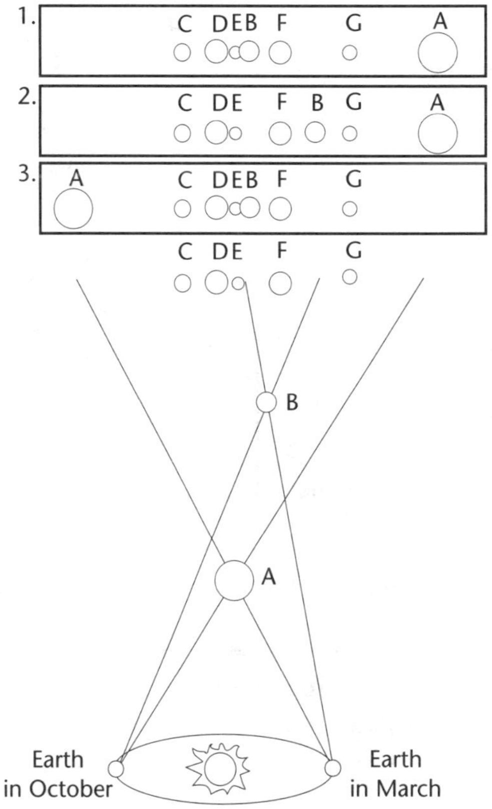


**21. Interpreting Diagrams** Some stars seem to be in slightly different positions when viewed from Earth in March compared to their positions in October. What is this difference called?

______________________________________

______________________________________

**22. Applying Concepts** Which of the diagrams at the top of the above figure shows the view of the stars from Earth in October?

______________________________________

______________________________________

**23. Inferring** Why does star A show a greater shift of position than star B?

______________________________________

______________________________________

## ◆ Essay

*Write a brief paragraph to answer each of the following. Use the back of this sheet or a separate sheet of paper if you need more room.*

**24.** Explain how an astronomer uses a spectrograph to study the stars.

______________________________________

______________________________________

______________________________________

______________________________________

**25.** Define absolute magnitude and apparent magnitude. Why do some stars with absolute magnitudes much greater than the sun's have much smaller apparent magnitudes?

______________________________________

______________________________________

______________________________________

______________________________________

## ◆ Using Science Skills

*Use the diagram below to answer questions 26–28. Use the back of this page or a separate sheet of paper if you need more room.*

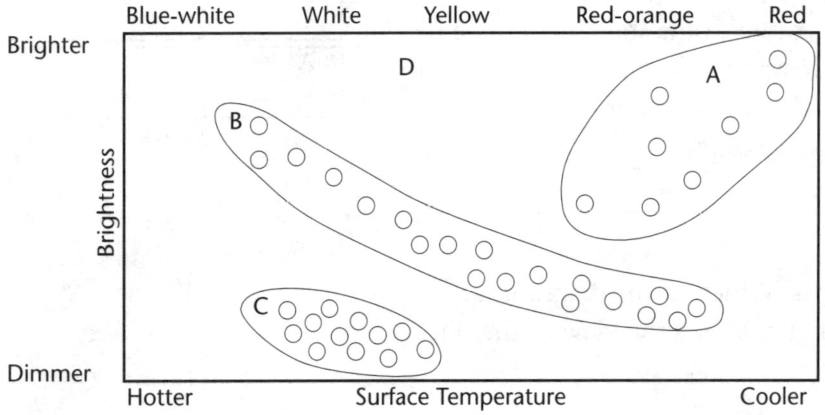


**26.** What is the name of the kind of diagram shown above?

______________________________________________

**27. Interpreting Diagrams** What is the letter of the region in which an average star like the sun is placed? What is the name of that region?

______________________________________________

**28. Interpreting Diagrams** Which region contains the brightest stars? What size stars would be in this region?

______________________________________________

______________________________________________

______________________________________________

## ◆ Essay

*Write a paragraph to answer each of the following. Use the back of this page or a separate sheet of paper if you need more room.*

**29.** How do astronomers think our solar system formed?

______________________________________________

______________________________________________

______________________________________________

**30.** Compare and contrast the three main kinds of galaxies.

______________________________________________

______________________________________________

______________________________________________

Name ______________________ Date __________ Class ______________



# Astronomy

## ◆ Multiple Choice

*Write the letter of the correct answer on the line at the left.*

______ **1.** You will see a total solar eclipse on Earth if you are within the
**a.** umbra. **b.** penumbra.
**c.** equinox. **d.** maria.

______ **2.** Satellites and space stations are used for all of the following except
**a.** communications. **b.** navigation.
**c.** research. **d.** controlling weather.

______ **3.** Scientists believe the moon has been hit by numerous meteoroids because they can see that it has
**a.** maria. **b.** rocks.
**c.** valleys. **d.** craters.

______ **4.** In a heliocentric system,
**a.** all the planets revolve around the sun.
**b.** the sun revolves around Earth.
**c.** the sun is the center of the universe.
**d.** the universe started with the big bang.

______ **5.** A streak of light made by a tiny rock passing through Earth's atmosphere is called a
**a.** asteroid. **b.** meteor.
**c.** meteorite. **d.** meteoroid.

______ **6.** The category of electromagnetic waves that any warm object gives off is called
**a.** x-ray **b.** radio
**c.** infrared **d.** ultraviolet

______ **7.** A light-year is
**a.** a star's magnitude compared to our sun's
**b.** a measure of speed.
**c.** the distance light travels in a year.
**d.** the distance from Earth to the sun.

______ **8.** The main characteristics used to classify stars are
**a.** size, temperature, and brightness.
**b.** size, distance from Earth, and temperature.
**c.** temperature, color, and brightness.
**d.** size, age, and brightness.

Name ______________________ Dat[e] ________ Class ______________

## BOOK TEST *(continued)*

______ **9.** A star with gravity so strong that even light cannot escape is a
- **a.** white dwarf.
- **b.** black hole.
- **c.** black dwarf.
- **d.** supernova.

______ **10.** A group of two or more stars is called a
- **a.** solar system.
- **b.** neutron star.
- **c.** protostar.
- **d.** star system.

## ◆ Completion

*Fill in the line to complete each statement.*

**11.** Day and night are caused by Earth's ______________ on its axis.

**12.** A satellite in ______________ orbit revolves around Earth at the same rate that Earth rotates.

**13.** The sun gets its energy from nuclear ______________ in its core

**14.** Astronomers classify galaxies as elliptical, irregular, and ______________.

**15.** When a star runs out of fuel, it becomes a ______________, a neutron star, or a black hole.

## ◆ True or False

*If the statement is true, write true. If it is false, change the underlined word or words to make the statement true.*

______________ **16.** The length of a star's life depends mainly on its <u>magnitude</u>.

______________ **17.** Astronomers use <u>parallax</u> to measure the distance to nearby stars.

______________ **18.** Two kinds of visible light telescopes are the <u>radio</u> telescope and the reflecting telescope.

______________ **19.** A star is born when the <u>expanding</u> gas and dust become so hot that nuclear fusion begins.

______________ **20.** According to the big bang theory, the universe formed about 10 to 15 <u>million</u> years ago.

**BOOK TEST** *(continued)*

## ◆ Using Science Skills

*Use the diagram below to answer the following questions.*

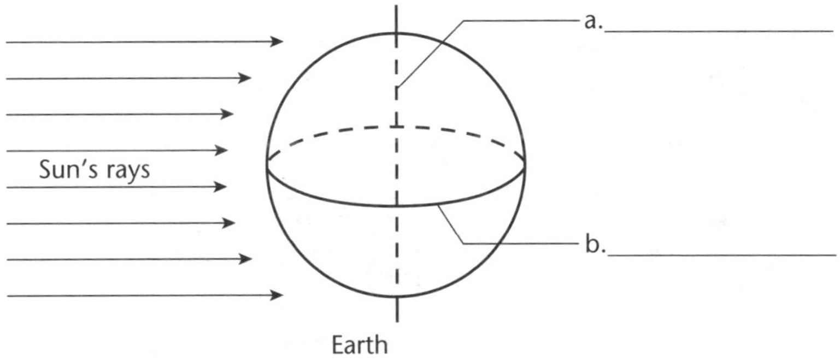


**21. Apply Concepts** Label Earth's equator and axis on the diagram.

**22. Interpret Illustrations** In which two seasons is Earth in the position shown? Explain.

______________________________________________

______________________________________________

## ◆ Essay

*Write an answer for each of the following.*

**23.** Why do you never see the far side of the moon from Earth?

______________________________________________

______________________________________________

**24.** How is Pluto different from the gas giants?

______________________________________________

______________________________________________

**25.** Compare and contrast the causes of solar eclipses and lunar eclipses.

______________________________________________

______________________________________________

**BOOK TEST** *(continued)*

## ◆ Using Science Skills

*Use the data in the table below to answer the following questions.*

| Planet | Diameter (kilometers) | Period of Rotation (Earth days) | Average Distance From the Sun (kilometers) |
|---|---|---|---|
| X | 4,878 | 59 | 58,000,000 |
| Venus | 12,104 | 243 | 108,000,000 |
| Earth | 12,756 | 1 | 150,000,000 |
| Mars | 6,794 | 1.03 | 228,000,000 |
| Jupiter | 142,800 | 0.41 | 778,000,000 |

**26. Interpreting Data** Which planet is planet X? How do you know?

______________________________________________

______________________________________________

**27. Interpreting Data** What is the approximate length of night on Jupiter in hours?

______________________________________________

______________________________________________

______________________________________________

## ◆ Essay

*Write an answer for each of the following.*

**28.** Briefly describe the life of a typical star like the sun.

______________________________________________

**29.** What does a spectrograph do? What useful information about stars can astronomers determine using a spectrograph?

______________________________________________

______________________________________________

______________________________________________

**30.** Compare and contrast the terrestrial planets with the gas giants.

______________________________________________

______________________________________________

______________________________________________

# Chapter 1

### Section 1-1 Review and Reinforce (p. 15)

1. winter
2. At point A the sun would be directly overhead, at point B it would be on the horizon, and at point C it would not be visible because it is night.
3. sunset; Earth rotates from west to east.
4. d
5. i
6. g
7. f
8. a
9. j
10. e
11. h
12. c
13. b

### Section 1-1 Enrich (p. 16)

1. Answers will vary. Students will probably find that telling time within 15 minutes is difficult with a sundial.
2. Answers will vary. Sample: A "moondial" couldn't be made because the moon's position in the sky from day to day varies much more than the position of the sun.
3. Accept any two: Sundials can't easily be moved. They don't work at night, indoors, or in cloudy weather. They are not as accurate as clocks.

### Section 1-2 Review and Reinforce (p. 19)

1. A new, B crescent, C first quarter, D full, E gibbous, F third quarter
2. Spring tides would occur when the moon is at A and D. Neap tides would occur when the moon is at C and F.
3. neap tide
4. eclipse
5. penumbra
6. tides
7. umbra
8. phase
9. spring
19. lunar
11. solar
12. gravity

### Section 1-2 Enrich (p. 20)

1. a lunar year is $12 \times 29.5 = 354$ days
   a solar year = 365 days
2. $19 \times 365 = 6{,}935$ days
   $19 \times 354 + (7 \times 29.5) = 6{,}932.5$ days
   No, it's off by 2.5 days.
3. The planting and growth of crops and the movements of animals are seasonal, so it is important for hunters and farmers to be able to accurately predict the seasons.

### Section 1-3 Review and Reinforce (p. 23)

1. A rocket moves forward when gases expelled from the rear push the rocket forward.
2. A single-stage rocket is a single unit. In a multistage rocket, up to three stages burn then drop away.
3. Satellites and space stations are used for communication, navigation, collecting weather data, and research.
4. Most rockets can only be used once. Space shuttles can be used many times.
5. A satellite in a geosynchronous orbit revolves around Earth at the same rate that Earth rotates. It seems to remain stationary over one point on Earth.
6. A satellite is any natural or artificial object that revolves around an object in space.

### Section 1-3 Enrich (p. 24)

1. The path is a result of the combination of the movement of the satellite and the movement of Earth.
2. Relative to any point on Earth, a satellite in geosynchronous orbit does not move, so it can't trace a path.
3. It would be the same distance to the west of 'Y' that 'Y' is from 'X.'
4. Every twelve hours.

5. The satellite should be in a geosynchronous orbit from which both the United States and Europe would be visible. The satellite would stay in the same relative position in the sky and could transmit from the United States to Europe 24 hours a day.
6. The satellite should be put into polar orbit, because it would eventually pass over the entire surface of Earth. A satellite in geosynchronous orbit only "sees" one side of Earth.

### Section 1-4 Review and Reinforce (p. 27)

1. The moon's diameter is about 1/4 that of Earth. It's mass is much less than Earth, about one-eightieth.
2. When the Earth was very young, an object about the size of Mars collided with it. Material from this collision was thrown into orbit around the Earth. Eventually, the materials combined to form the moon.
3. Galileo. He saw craters, highlands, and maria with it.
4. A Galileo observes the moon with a telescope, B President Kennedy announces American space flight program, C First humans land on moon
5. The *Apollo* program obtained information about the composition of moon rocks, about moonquakes, and about heat flow from the moon.
6. He put two lenses in a wooden tube.
7. Craters are pits on the moons surface that are caused by the impact of meteorites.
8. Maria are flat, dark parts of the moon's surface formed by the flow of molten rock billions of years ago.

### Section 1-4 Enrich (p. 28)

1. Region D is probably the oldest, and region A is probably the youngest.
2. Crater 2 is older.
3. Crater 3 is probably older, because many more meteors have struck it than have struck crater 2.
4. Crater 4 must be older than region A, because the material of region A could not have filled its floor if it hadn't already been there.
5. It must be younger. Parts of the region's floor couldn't collapse if the region hadn't already been there. From oldest to youngest they are: crater 4, the floor of region A, the labeled rill, crater 1.

### Chapter 1 Skills Lab (pp. 29–31)

For answers, see Teacher's Edition, pp. 22–23.

### Chapter 1 Skills Lab (pp. 32–33)

For answers, see Teacher's Edition, pp. 30–31.

## Chapter 2

### Chapter 2 Project Worksheet 1 (p. 38)

**Distance from Sun**

**1:10,000,000,000 Scale:** Sun Center; Mercury 5.8 m; Venus 10.8 m; Earth 15.0 m; Mars 22.8 m; Jupiter 77.8 m; Saturn 142.7 m; Uranus 287.1 m; Neptune 449.7 m; Pluto 591.3 m

**1:50,000,000,000 Scale:** Sun Center; Mercury 1.2 m; Venus 2.2 m; Earth 3.0 m; Mars 4.6 m; Jupiter 15.6 m; Saturn 28.5 m; Uranus 57.4 m; Neptune 89.9 m; Pluto 118.3 m

**Object Diameter**

**1:1,000,000,000 Scale:** Sun 1.39 m; Mercury 5 mm; Venus 12 mm; Earth 13 mm; Mars 7 mm; Jupiter 14.3 cm; Saturn 12.1 cm; Uranus 5.1 cm; Neptune 5.0 cm; Pluto 2 mm

**1:500,000,000 Scale:** Sun 2.78 m; Mercury 10 mm; Venus 24 mm; Earth 26 mm; Mars 14 mm; Jupiter 28.6 cm; Saturn 24.1 cm; Uranus 10.2 cm; Neptune 9.9 cm; Pluto 4 mm

**Model Objects:** These will vary but should be about the diameters computed for the scales chosen by students.

### *Chapter 2 Project Worksheet 2 (p. 39)*

**Combination Model**

**Diameter of Objects:** Sun 13.9 cm; Mercury 0.5 mm; Venus 1.2 mm; Earth 1.3 mm; Mars 0.7 mm; Jupiter 14.3 mm; Saturn 12.1 mm; Uranus 5.1 mm; Neptune 5.0 mm; Pluto 0.2 mm

**Distance from Sun:** Sun Center; Mercury 5.8 m; Venus 10.8 m; Earth 15.0 m; Mars 22.8 m; Jupiter 77.8 m; Saturn 142.7 m; Uranus 287.1 m; Neptune 449.7 m; Pluto 591.3 m

**Model Objects:** These will vary, but they should be about the size indicated in Diameter of Objects, above.

**Analyzing and Presenting**

1. Students should write a paragraph on their experiences, including information on the scales that they used, why they chose those scales, and what problems they encountered.
2. Students should write a paragraph on the distances between the planets and the sun, discussing how the solar system is largely empty space. They should mention that the spacing between the planets is not equal—some planets are closer together than are others.
3. Students should write a paragraph about scaling. One of the advantages to using scale models is that it allows you to work with measurements that are too large to visualize easily. A disadvantage is that it is difficult to chose a scale that is convenient for showing both large distances and relatively small diameters in one model.

### *Section 2-1 Review and Reinforce (p. 43)*

1. In the geocentric model, the planets and the sun revolve around Earth. In the heliocentric model, Earth and the planets revolve around the sun.
2. Ptolemy was trying to explain why the planets seem to move at different speeds and sometimes appear to move backwards among the stars.
3. Galileo observed moons orbiting around Jupiter. This showed that not everything revolves around Earth. He also observed phases of Venus that are similar to Earth's moon. The heliocentric model could easily explain these observations.
4. The sun's gravity pulls the planets toward the sun while inertia keeps them moving ahead. The result is that the planets orbit the sun.
5. heliocentric
6. ellipse
7. inertia
8. geocentric

### *Section 2-1 Enrich (p. 44)*

1. The full moon is on the opposite side of Earth from the sun. The full Venus is in almost the same direction as the sun. The apparent size of the moon does not change with its phase.
2. The new moon is between Earth and the sun. So is the new Venus.
3. Because its size does not appear to change, we know the moon is always about the same distance from Earth. Also, the moon is sometimes between Earth and sun, and sometimes on the opposite side of Earth from the sun.

### *Section 2-2 Review and Reinforce (p. 47)*

1. chromosphere
2. photosphere
3. core
4. corona
5. sunspot
6. prominence
7. f
8. h
9. b
10. d
11. c
12. i
13. a
14. g
15. e

### Section 2-2 Enrich (p. 48)

1. Astronomers inferred that the sun rotates because sunspots move in one direction over time.
2. The sun rotates from west to east.
3. Sunspots at points A and B take longer to move around the sun than does a sunspot on the equator.
4. Different parts of the sun's surface rotate at different rates, so scientists must give a range for the time of the sun's rotation.

### Section 2-3 Review and Reinforce (p. 51)

1. Mercury
2. Venus
3. Earth
4. Mars
5. Mercury, Venus, Earth, Mars
6. Earth
7. Venus
8. Mars
9. Earth, Mars
10. Venus, Earth
11. Mercury
12. Venus
13. Earth, Mars
14. Mars
15. The first four planets from the sun: Mercury, Venus, Earth, and Mars. They are the earthlike planets with rocky surfaces.
16. Rotation from east to west, in the direction opposite from the usual direction
17. The trapping of heat by the atmosphere

### Section 2-3 Enrich (p. 52)

1. Carbon dioxide either becomes dissolved in the oceans or is absorbed by green plants.
2. On Earth, at any one time, most of the carbon dioxide is locked up in rocks. On Venus, it is in the atmosphere.
3. On Earth, carbon dioxide dissolves in ocean water. While there, it reacts with other chemicals to form rocks. On Venus, there are no oceans in which the carbon dioxide can dissolve.
4. If Earth's temperature increased by 100°C, the amount of carbon dioxide in the atmosphere would increase. This is because the oceans would evaporate and green plants would not be able to live. Without these, the carbon dioxide would have no way out of the atmosphere.
5. The carbon dioxide would create a greenhouse effect that would further warm Earth's surface.

### Section 2-4 Review and Reinforce (p. 55)

1. Jupiter, Saturn, Uranus, Neptune, Pluto
2. Jupiter, Saturn, Uranus, Neptune
3. The gas giants are much larger than Pluto. The gas giants are made mostly of gas but Pluto is solid.
4. The gas can't escape because gravity is so great.
5. the sun
6. Jupiter
7. ice and rocks
8. Answers may vary. Students' answers should refer to the fact that Neptune's orbit was predicted by mathematicians who calulated where it must be based on its effect on the orbit of Uranus.
9. They are nearly the same size.
10. A gas giant is a large planet made almost entirely of gas. The gas giants are Jupiter, Saturn, Uranus, and Neptune

### Section 2-4 Enrich (p. 56)

1. 29.7 a.u.
2. 30 a.u.
3. In 2228
4. In 2113
5. Pluto (after February 1999)

### *Section 2-5 Review and Reinforce (p. 59)*

**Asteroid**
Description: Object revolving around the sun that is too small to be considered a planet
Location/Movement: Revolves around the sun between the orbits of Mars and Jupiter

**Comet**
Description: Chunk of ice and dust with a bright head and long tail
Location/Movement: Orbits the sun in a long narrow ellipse

**Meteoroid**
Description: Chunk of rock or dust, usually from a comet or asteroid
Location/Movement: Moves through the solar system. If formed from comet would have similar elliptical orbit

1. A meteoroid enters Earth's atmosphere and friction causes it to burn. If the meteoroid is big enough, it does not completely burn and it hits Earth's surface.
2. The head is the main body of the comet. The nucleus of the head is made of ice and dust. The coma is the halo of gas and dust, formed when sunlight turns the ice into gas. The tail is the gas from the comet that is pushed away from the sun by the solar wind.
3. Comets are visible for days or weeks. A meteor is a quick flash of light that moves across the sky.
4. meteor
5. comet
6. meteorite
7. asteroid
8. meteoroid
9. asteroid belt

### *Section 2-5 Enrich (p. 60)*

1. Craters on the moon were caused by impacts of large meteoroids. The moon has been revolving around Earth for a long time. Since large meteoroids have hit the moon many times in the past, they have probably hit Earth, too.
2. Many meteoroids never hit Earth's surface because they burn up in Earth's atmosphere. When a meteoroid does hit the surface, wind and rain often destroy the crater.
3. About 65 million years old
4. Deep below Earth's surface, rocks from space
5. A large meteoroid might have hit Earth 65 million years ago, causing meteor dust to be spread around Earth. This is one possible explanation of the mass extinction of dinosaurs and other creatures at that time.

### *Section 2-6 Review and Reinforce (p. 63)*

1. Students should list three of the four conditions given: All living things are made up of one or more cells, take in energy and use it to grow and develop, reproduce, and give off waste.
2. "Goldilocks conditions" are the "just right" conditions on Earth that scientists believe are necessary for life.
3. The presence of liquid water, a suitable temperature range, and a suitable atmosphere
4. Perhaps the "Goldilocks conditions" aren't necessary for life.
5. The channels may have been formed by flowing water. Since life as we know it requires liquid water, scientists hypothesize that Mars may have once had the conditions needed for life to exist.
6. Europa is covered with an icy crust that has similar patterns to the ice crust over Earth's Arctic Ocean. Some scientists think this may indicate that there is liquid water under the ice. If there is liquid water on Europa, there might also be life.
7. Extraterrestrial life would be life other than that on Earth.

### Section 2-6 Enrich (p. 64)

1. Planet A could not support human life. Its gravity is too great.
2. Planet B might be able to support human life. Its mass is over 0.4 times Earth's, so it might have water and an atmosphere. Its length of day is close to Earth's. Its gravity is slightly less than Earth's so humans wouldn't be harmed.
3. Planet C could not support human life. Its day is too long.
4. Even though Venus's mass is very similar to Earth, it could not support human life if it were placed in Earth's orbit because its day is far too long.
5. Even though its day is almost the same length as Earth's, Mars could not support human life because it is too small to have a thick enough atmosphere or enough surface water.

### Chapter 2 Real-World Lab (p. 65–66)

For answers, see Teacher's Edition, p. 65.

### Chapter 2 Skills Lab (p. 67–69)

For answers, see Teacher's Edition, pp. 58–59.

## Chapter 3

### Section 3-1 Review and Reinforce (p. 79)

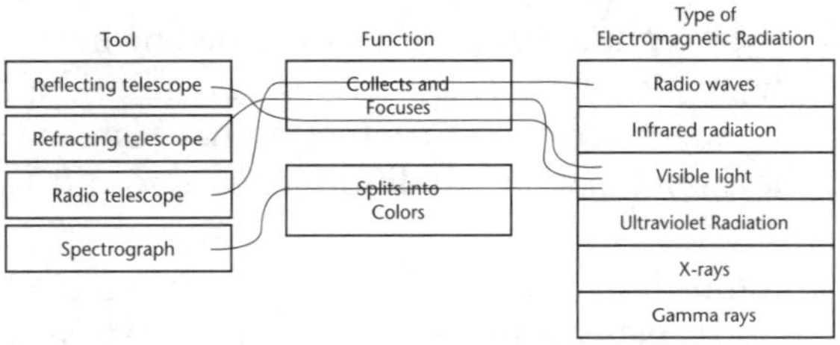


1. surface temperature of stars, chemical composition of stars
2. d
3. c
4. a
5. f
6. b
7. e

### Section 3-1 Enrich (p. 80)

1. Gamma radiation should give the highest resolution image, because it has the shortest wavelength.
2. Answers may vary. Sample: Satellites are expensive to build and launch. A large telescope, with high resolution, would be especially expensive to launch. If a telescope breaks in space, it is harder to fix than if it were on the ground.
3. Radio waves, because they could be observed more cheaply and easily from ground-based radio telescopes. Although they would give a low-resolution image, this would be okay because the astronomers are looking at broad areas of the sky, not tiny details.
4. Infrared radiation, because only it can pass through the clouds of gas and dust in space and because its wavelength is short enough to create a higher resolution image.

### Section 3-2 Review and Reinforce (p. 83)

1. A
2. F
3. Their distance from Earth
4. Astronomers use parallax to measure the distance to nearby stars. They measure how much the star appears to move against a background of stars that are much farther away.
5. The main characteristics used to classify stars are size, temperature, and brightness.
6. The sun is a very bright, medium-sized star, about 6,000 degrees Celsius.
7. absolute magnitude
8. galaxy
9. light-year
10. main sequence
11. Hertzsprung-Russell diagram
12. Parallax
13. universe
14. apparent magnitude
15. giant star

### Section 3-2 Enrich (p. 84)

1. The difference in magnitude is 1. Procyon is about 2.5 times brighter than Regulus.
2. Venus' magnitude is 2 less than Mars, so Venus is about 2.5 × 2.5, or about 6.3 times brighter than Mars.
3. The difference in magnitude between the dimmest star that can be seen with the naked eye, and the dimmest one that can be seen with binoculars is 4. This means that the dimmest naked-eye star is about 2.5 × 2.5 × 2.5 × 2.5 = about 39 times brighter than the dimmest binocular star.

### Section 3-3 Review and Reinforce (p. 87)

1. D
2. B
3. A
4. E
5. C
6. mass
7. e
8. c
9. b
10. d
11. h
12. g
13. a
14. f

### Section 3-3 Enrich (p. 88)

1. 32 mm
2. 8 mm (4 mm + 4 mm)
3. You are looking through more of the shell along line 3.
4. The shell would appear thicker and brighter along line 3.
5. The shell must be glowing, since it shows up as a light color. When you look through the outer part of the shell, your line of sight includes more of the glowing shell of gas than when you look nearer the center, so you see a ring.

### Section 3-4 Review and Reinforce (p. 91)

1. B
2. A triple star
3. The position of the visible star might wobble slightly as the heavy unseen star revolves around it.
4. A group of two or more stars revolving around each other
5. A star system made up of two stars
6. A binary star system in which a brighter star appears to dim when the dimmer star of the pair passes in front of it
7. Spiral galaxies are galaxies with spiral arms radiating out from the center. They contain lots of dust and gas.
8. Elliptical galaxies look like flattened balls and do not contain lots of dust and gas.
9. Irregular galaxies are galaxies with irregular shapes.

### Section 3-4 Enrich (p. 92)

| | Period | Brightness | Distance |
|---|---|---|---|
| **A** | 5 | 1,250 | 500 |
| **B** | 20 | 3,000 | 700 |
| **C** | 35 | 5,000 | 1,000 |
| **D** | 50 | 7,500 | 1,200 |
| **E** | 90 | 12,500 | 1,600 |

Note: Since students will be using rulers to read the graph, their answers will be approximations only.

### Section 3-5 Review and Reinforce (p. 95)

1. Away from Earth
2. The farther away a galaxy is, the faster it moves away from Earth.
3. 4
4. 5
5. 1
6. 7
7. 3
8. 2
9. 6

**10.** The big bang theory is the theory that the universe formed in an huge explosion 10 to 15 billion years ago.

### *Section 3-5 Enrich (p. 96)*

| | Rank |
|---|---|
| Galaxy A | 1 |
| Galaxy B | 4 |
| Galaxy C | 2 |
| Galaxy D | 3 |

**1.** Galaxy B is farthest from Earth.
**2.** No, because none of the spectrums is blue shifted.

### *Chapter 3 Real-World Lab (pp. 97–98)*

For answers, see Teacher's Edition, p. 101.

### *Chapter 3 Skills Lab (pp. 99–101)*

For answers, see Teacher's Edition, p. 110.

## Interdisciplinary Explorations

### *Language Arts: Ain't I a Woman (p. 102)*

**1.** Women need to be helped to do everyday things.
**2.** She challenges his generalization about women by listing the things she has achieved alone.

### *Language Arts: Words from Mars Fiction (p. 103)*

**1.** scrutinize
**2.** fancied
**3.** complacency
**4.** infusoria
**5.** serene
**6.** transient

### *Language Arts: Move to Mars (p. 104)*

Accept all reasonable advertisements. Make sure students have responded to all of the suggestions and answered all of the questions.

### *Science: Gravity on Mars (p. 105)*

**1.** 780 N
**2.** 120 Newtons
**3.** 320 Newtons
**4.** 20 Newtons
**5.** on Earth: 490 N; on Mars: 200 N
**6.** Gravity on Earth is about two and a half times stronger than on Mars.
**7.** Answers will vary. Heights on Mars two to three times those on Earth are the most reasonable.

### *Science: The Polar Caps of Mars (p. 106)*

**1.** They assumed that both polar caps had the same chemical composition.
**2.** From Earth. In 1948, there were no spacecraft yet in space.
**3.** They did not know that Earth's atmosphere was absorbing some spectral waves.
**4.** Since the scientists concluded the caps were carbon dioxide, it can be inferred that the models for water ice did not match the observed growth and decline of the caps on Mars.
**5.** The northern polar cap contains a core of water ice that never melts. The southern polar cap does not contain water ice, so it copmpletely disappears.
**6.** The north polar cap because the colonists can use the water.

### *Science: History of Mars Exploration (p. 107)*

Students' timelines will vary. Accept all accurate and researched timelines.

1925 First launch of liquid fuel rocket

1958 NASA created

1962 *Mariner 2* flyby

1965 *Mariner 5* flyby

1964 *Mariner 4* photographs

1969 *Mariner 6* and *7* photographs

1971 Soviet *Mars 3* first landing;

1971 *Mariner 9* orbits

1974 Soviet *Mars 5*

1976 *Vikings 1* and *2* land

1996 *Mars Global Surveyor*

1997 *Mars Pathfinder*

### *Social Studies: Earth from Space (p. 108)*

Accept all reasonable, researched findings. Make sure students have responded to all of the suggestions and answered all of the questions.

1. Hurricanes, thunderstorms, flashes of light
2. Yes, brown haze
3. Massive, only smoke
4. Smoke plume
5. Few artificial features are visible from space
6. Major cities such as Tokyo or New York, oil flares in Persian Gulf

Students who have Internet access can find thousands of images of Earth with accompanying text on NASA's "Earth from Space" website at **earth.jsc.nasa.gov**.

### *Social Studies: The Search for Life on Mars (p. 109)*

Accept all researched timelines. Students with access to the Internet can begin their research by accessing Syracuse University's Physics Department website of SETI. SETI is an organization committed to searching for extra-terrestrial intelligent life. The site is at **www.phy.syr.edu/courses/modules/SETI**.

### *Social Studies: Map of Mars (p. 110)*

Accept all neatly drawn, researched maps.

Students who have Internet access can find some of the newest information available on Mars from NASA's Mars site at **pds.jpl.nasa.gov/planets/welcome/mars.htm**.

### *Mathematics: A Year on Mars (p. 111)*

Accept all calendars that have 669 sols and the correct proportion of days in each season for the northern and southern hemispheres.

### *Mathematics: Model the Mars Neighborhood (p. 112)*

1. Students' answers may vary slightly depending on the source they used.

| | | |
|---|---|---|
| Earth | 150 million km from sun | 12,756 km diameter |
| Mars | 228 million km from sun | 6,787 km diameter |
| Venus | 108 million km from sun | 12,104 km diameter |
| Moon | 384,400 km from Earth | 3,476 km diameter |
| Phobos | 9,378 km from Mars | about 21 km diameter |
| Deimos | 23,459 km from Mars | about 12 km diameter |

2. smallest: about 9 cm; largest: about 2.2 km
3. Answers will vary. Make sure students' answers are consistent with the scale they choose.
4. smallest: about 0.1 cm; largest: about 127 cm
5. Answers will vary. Make sure students' answers are consistent with the scale they choose.
6. Check that students' models are consistent with the scale they chose.

### *Mathematics: Martian Seasons in Days (p. 113)*

| | Northern Hemisphere (Earth days) | Northern Hemisphere (Earth months and days) |
|---|---|---|
| **Winter** | 158 | 5 months 8 days |
| **Spring** | 199 | 6 months 19 days |
| **Summer** | 185 | 6 months 5 days |
| **Fall** | 144 | 4 months 24 days |

| | Southern Hemisphere (Earth days) | Southern Hemisphere (Earth months and days) |
|---|---|---|
| **Winter** | 185 | 6 months 5 days |
| **Spring** | 144 | 4 months 24 days |
| **Summer** | 158 | 5 months 8 days |
| **Fall** | 199 | 6 months 19 days |

## Assessment

### *Chapter 1 Performance Assessment (p. 116)*

**Analyze and Conclude**

1. Students should show two positions of Earth, the sun, and the moon in a line, with the sun farthest from Earth and the moon. In one position, Earth is between the sun and the moon. In the other position, the moon is between Earth and the sun. In both positions, a high tide is on the side of Earth facing the moon, and on the side facing away from the moon.
2. When the moon is on the opposite side of Earth from the sun, it is full. When the moon on the side toward the sun, it is new.
3. If connected by a line, the positions of Earth, the sun, and the moon should form a right triangle, with the sun at the vertex of the two longest sides. The moon will be at either side of Earth in the two positions. In both positions, a high tide is on the side of Earth facing the moon, and on the side facing away from the moon.
4. In both positions, the moon will be half illuminated by the sun. The two positions are called first quarter and third quarter.

### *Chapter 1 Test (p. 117–120)*

1. c
2. d
3. b
4. b
5. c
6. a
7. d
8. c
9. b
10. c
11. solstices
12. penumbra
13. geosynchronous
14. full
15. equinox
16. rotations
17. vernal equinox
18. lunar eclipse
19. true
20. true
21.
   **A.** overhead
   **B.** on the horizon
   **C.** on the horizon
   **D.** not visible
22.
   **E.** first quarter
   **F.** new moon
   **G.** last quarter
   **H.** full moon
23. At most points by the seashore there are two high tides separated by two low tides during 25 hours. The length of time between a high and low tide is more than 6 hours.

**24.** The seasons are caused by the tilt of Earth's axis in relation to the plane of the orbit and the revolution of Earth around the sun. When the Northern Hemisphere is pointed toward the sun, it is warm because the sun hits the Northern Hemisphere most directly. We call that summer. Six months later, sunlight hitting the Northern Hemisphere is at the greatest angle, and provides less heat, so there is winter. During spring and fall sunlight hits at an angle between the maximum and minimum, and the Northern Hemisphere has spring and fall.

**25.** About 4.5 billion years ago, an object about the size of Mars collided with Earth. Material from the object and from Earth's outer layers was thrown into orbit around Earth. Eventually, this material formed the moon.

**26.** Neap tides—positions B and D because the gravity of the moon and sun are pulling at right angles. Spring tides—positions C and E, because the gravity of the sun and moon would pull along in the same line.

**27.** During spring tides, the moon would be either new or full. A person standing at position A would observe a low tide.

**28.** The back side of the moon is not visible from Earth because it always faces away. Astronomers would need to send satellites to see the back side of the moon or astronauts could go there.

**29.** Each month, as the moon revolves around Earth, it crosses between Earth and the sun. Occasionally, it crosses directly between Earth and the sun. Then the moon's shadow hits Earth's surface. Anyone in this shadow sees a solar eclipse. Occasionally, the moon passes directly behind Earth, so that the moon enters Earth's shadow. People on Earth then observe a lunar eclipse. Lunar eclipses are more common than solar eclipses because Earth is larger than the moon and makes a wider shadow.

**30.** Answers will vary. Sample: The Apollo astronauts brought instruments such as seismometers to study moonquakes and another instrument to study the moon's interior. They learned about the moon's soil by observing their footprints, and they made observations by traveling in a moon buggy. Analysis of rocks brought back to Earth showed that most of the rocks had been molten, so the surface must have been very hot. Other rocks showed evidence that meteors had hit the moon.

## *Chapter 2 Performance Assessment (p. 123)*

### Analyze and Conclude

**1.** Answers will vary. Samples:

**The sun.** Contains over 99% of all of the mass of our solar system. The sun's energy comes from nuclear fusion. The sun has an atmosphere with three parts. From inside to out they are: the photosphere, which is where the sun's light comes from; the chromosphere; and the corona. The chromosphere and corona are visible only during eclipses. Features on the sun include the following: Sunspots are dark, cooler areas on the sun; prominences are loops connecting sunspots; solar flares are huge explosive loops of gas.

**Mercury.** The smallest of the inner planets and closest to the sun. No moons. Rotates on its axis once each 59 days. Has almost no atmosphere, very hot on the side facing the sun but very cold on the other side.

**Jupiter.** One of the "gas giants." Largest planet. Its day is less than 12 hours long. Its Great Red Spot is apparently a kind of storm in its atmosphere. 17 known moons, four of which were discovered by Galileo. One of its moons, Europa, may have liquid water beneath an icy crust.

**Comets.** Most comets have long, elliptical orbits that take them close to the sun and back out far beyond Earth's orbit. When heated by the sun comets have three parts: the nucleus or solid that is something like a dirty snowball; a coma formed by dust released when sunlight turns the ice into gas; and a long tail that can be hundreds of millions of kilometers long. The tail always points away from the sun, even when the comet itself is moving away from the sun.

2. Answers may vary. Samples: The relative size of the planets is difficult to show so that wasn't included. For example, I wanted to draw each of my objects so that anyone using my cards could compare the sizes, but even if I drew the sun as large as possible on its card, a tiny dot would have been too big to represent the comet or Mercury. The relative distances of the objects from the sun are also difficult to show.

### Chapter 2 Test (pp. 124–127)

1. c
2. d
3. b
4. d
5. a
6. c
7. b
8. d
9. a
10. b
11. ellipses/elliptical
12. Sunspots
13. Meteorites
14. water
15. comets
16. inertia
17. heliocentric
18. true
19. true
20. fusion
21. Planet C will take the longest amount of time to revolve around the star because it is the greatest distance from the star.
22. Planet B will have the greatest gravitational pull because it has the largest mass. The greater the mass, the greater its gravitational pull.
23. Gravity pulls planets toward the sun while the planets' inertia keeps them moving forward.
24. Nuclear fusion occurs in the extreme heat and pressure at the sun's core. Hydrogen atoms join to form helium atoms. The helium atoms have a lesser mass than the total mass of the hydrogen from which they are formed. The lost mass is changed into energy.
25. Jupiter, Saturn, Uranus, and Neptune, the gas giants, are all relatively large in size, and they are made almost entirely of gas. The inner planets are smaller and solid.

**26.–27.**

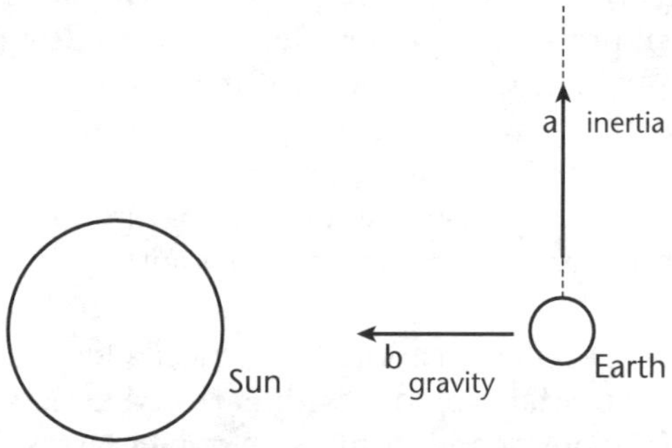


28. The side of Mercury that faces the sun is extremely hot because it is so close to the sun. Since Mercury has almost no atmosphere, at night all the heat escapes into space.
29. Pictures of Mars revealed crisscross patterns in the surface that might be old stream beds. If there was once liquid water on Mars, there may have been life.
30. Heat from the sun is held in by the thick atmosphere.

### Chapter 3 Performance Assessment (p. 130)

1. The small object in the model must represent the object from the second column of the table.
2. The large object in the model must represent the object from the first column of the table.
3. Answers may vary. Sample: There are large differences in size between the categories of objects discussed.
4. Answer may vary but should recognize that the third object would have to be 100 times smaller or larger.
5. The ratio of the largest to smallest object would be about 10,000,000 to 1. If an object 1 mm in diameter were chosen as the smallest, the largest would be about 10 kilometers in diameter.
6. Students' diagrams should be similar to "Exploring the Lives of Stars," on page 115 of the text. A star's mass determines its life history.

### *Chapter 3 Test (pp. 131–134)*

1. b
2. b
3. d
4. a
5. a
6. b
7. c
8. b
9. b
10. a
11. eclipsing binary
12. radio telescope
13. refracting
14. reflecting
15. nebula
16. true
17. true
18. true
19. apparent magnitude
20. true
21. parallax shift
22. view 2
23. It is closer than star B.
24. A spectrograph breaks the light from an object into colors and takes a photograph of the resulting spectrum. Spectrums from different stars vary, depending on the temperature of the stars. Astronomers examine the spectrums of stars to tell what chemicals are in the stars and how hot the stars are.
25. A star's absolute magnitude is how bright it would appear if it were at a standard distance from Earth. A star's apparent magnitude is how bright it actually appears from Earth. Stars with much greater absolute magnitudes than the sun have apparent magnitudes much smaller than the sun because they are much farther away from the sun.
26. Hertzsprung-Russell diagram, or H-R diagram
27. Region B. Main Sequence.
28. Region A contains the brightest stars. These stars would be giant and supergiant stars.
29. About 5 billion years ago, a nebula started collapsing. Gradually, it formed a spinning disk. When gravity pulled enough material into the center, the sun was born. Elsewhere in the disk, solid spheres formed. The spheres closest to the sun lost most of their atmospheres and became the inner planets. The spheres farthest from the sun became the gas giants. Outside the gas giants, a huge cloud formed. This cloud is probably the main source of comets. Pluto also formed in this region.
30. Galaxies are made of billions of stars. Spiral galaxies have long arms that spiral out from the center. Earth is in a spiral galaxy called the Milky Way. Spiral galaxies have large amounts of gas and dust. Elliptical galaxies look like flattened balls. They lack gas and dust, so new stars cannot form in them. Irregular galaxies have irregular shapes.

## Book J Test

***(pp. 137–140)***

**1.** a **2.** d
**3.** d **4.** a
**5.** b **6.** c
**7.** c **8.** a
**9.** b **10.** d
**11.** rotation
**12.** geosynchronous
**13.** fusion
**14.** spiral
**15.** white dwarf
**16.** mass
**17.** true
**18.** reflecting
**19.** contracting
**20.** billion
**21.** **a.** equator; **b.** axis
**22.** Spring and fall. The sun is directly overhead at the equator at noon on the vernal and autumnal equinoxes.
**23.** The moon revolves around Earth in the same time it takes to rotate on its axis. So the same side of the moon is always facing Earth.
**24.** Answers may vary. Sample: Unlike the gas giants, Pluto is small and rocky. It has only one moon but the gas giants have many. Pluto's moon, Charon, is so large compared to Pluto that some astronomers consider the pair to be a double planet.
25. In both kinds of eclipses the sun, moon, and Earth are in a line. In a solar eclipse, the moon is between Earth and the sun, so that the moon's shadow falls on Earth. In a lunar eclipse, Earth is between the sun and the moon, and Earth's shadow falls on the moon.
**26.** Planet X must be Mercury because Mercury is the only planet closer to the sun than Venus.
**27.** Night would be about 5 hours long ($0.41 \times 24 \div 2$).
**28.** The earliest stage of star's life is a protostar. Eventually, gravity causes gas and dust to contract and become hot enough for nuclear fusion to begin. A medium-sized star burns for millions of years, until it begins to run out of fuel. Then it turns into a red giant. Eventually, the outer parts of the star drift off into space, and what remains behind is a white dwarf.
**29.** A spectrograph is a device that divides light into its component colors. By looking at the resulting spectrum, astronomers can find out the temperatures and chemical compositions of stars.
**30.** The terrestrial planets are the first four from the sun—Mercury, Venus, Earth, and Mar. Compared to the gas giants, they are relatively small. All are rocky. Being closer to the sun, they receive a lot of sunlight, so their sunlit surfaces are warm or hot. The gas giants are the next four planets—Jupiter, Saturn, Uranus, and Neptune. As their name suggests, they are made up almost entirely of gases, and they are much larger than Earth.